Lecture Notes in Mobility

Series Editor

Gereon Meyer, VDI/VDE Innovation und Technik GmbH, Berlin, Germany

More information about this series at http://www.springer.com/series/11573

Jochen Langheim
Editor

Electronic Components and Systems for Automotive Applications

Proceedings of the 5th CESA Automotive Electronics Congress, Paris, 2018

Springer

Editor
Dr.-Ing. Jochen Langheim
STMicroelectronics
Paris, France

ISSN 2196-5544 ISSN 2196-5552 (electronic)
Lecture Notes in Mobility
ISBN 978-3-030-14158-5 ISBN 978-3-030-14156-1 (eBook)
https://doi.org/10.1007/978-3-030-14156-1

Library of Congress Control Number: 2019934796

This Springer imprint is published by the registered company Springer Nature Switzerland AG
The registered company address is: Gewerbestrasse 11, 6330 Cham, Switzerland

Technical Committee

Chairman

Jochen Langheim, STMicroelectronics

Program Committee

Ignacio Alvarez, Faurecia
Philippe Aumont, SIA
Patrick Bastard, Renault
Jean Luc Brossard, PFA
Jean-Philippe Dehaene, Vector
Kadiri El Khamis, Groupe PSA
Francois Fischer, ERTICO
Jean-Laurent Franchineau, VEDECOM
Pierre-Yves Geels, Segula Matra Automotive
Wolfgang Gessner, VDI/VDE
Hervé Gros, SIA
Anne Guillaume, LAB
Jochen Langheim, STMicroelectronics
Gilles Le Calvez, Vedecom
Pierre Lebrun, Valeo
Carlos Lee, Phontonics Cluster
Gereon Meyer, VDI/VDE
Jérôme Perrin, Renault
Vanessa Picron, Valeo
Ladimir Prince, Groupe PSA
Patrice Reilhac, VALEO
Pascal Ribot, STMicroelectronics

Daniel Richard, Valeo
Gillez Rizzo, ACSIEL
David Roine, Valeo
Wolfgang Runge, ELIV Baden Baden
Paul Schimmerling, SIA
Jean-François Sencerin, Renault
Andy Toulemonde, Infineon
Louis-Claude Vrignaud, Continental
Gérard Yahiaoui, Nexyad

Contents

Part I
Introduction

SIA CESA 2018—Electric Components and Systems for Automotive Applications

Jochen Langheim and Hervé Gros

Every second year the French Society of Automotive Engineers organizes in Paris the SIA-CESA conference on Electric Components and Systems for Automotive Applications. It addresses actual questions in the domain of all electronics components and systems with the aim to give a market insight and open the floor for business discussions.

In addition, some of the presenters contribute to, this book that highlights some of the most interesting topics.

Electronics took in the last decades a growing share of the value of a car. While at the end of the last century, its share in the production value was in the area of around 10%, it is today expected to rise beyond 50% in the next ten years. Driver assistance, car connectivity, electric traction or cloud experience with Artificial Intelligence (AI) & big data processing are among the most important drivers in this new era.

This revolution has of course important consequences on the way how to address mobility with new services, the competences needed in development departments to address new subjects, on the supply chain, which will see new actors replacing old ones. It also has an impact onto our society, about the way to deal with business on one hand based on data, personalized publicity and anticipation of maintenance, but also on the other hand on threats like loss of privacy, hacking and altering of application behaviors or misuse of new features for law infringement with criminal or even terrorist ideas behind.

Main goal of the SIA-CESA conference is to bring together communities from different areas, industrial, but also from legal, law enforcement or data protection. We

J. Langheim (✉)
STMicroelectronics, 29 Bd Romain Rolland, 75669 Paris, France
e-mail: jochen.langheim@st.com

H. Gros
Société des Ingénieurs Automobiles (SIA), 79 rue Jean-Jacques Rousseau, 92158 Suresnes, France
e-mail: herve.gros@sia.fr

J. Langheim (ed.), *Electronics Components and Systems for Automotive Applications*, Lecture Notes in Mobility,
https://doi.org/10.1007/978-3-030-14156-1_1

did succeed this time the challenge to motivate our colleagues from the engineering community to come to this event and meet with other domains then their own one.

This year, the automotive community was supported by the French Automobility Platform (PFA) and represented by the main R&D and Engineering leaders of the French Groups as well as representatives of the automotive industry from countries like Germany or Canada.

The electronics industry was represented in particular by the communities of the French Strategic Committee of the sector (CSF), IoT (AIOTI), smart systems (EPoSS and EURIPIDES) and photonics (EPIC).

Besides, CESA had the honor to welcome also the French National Police (Gendarmerie), the French Data Protection Office (CNIL) and representatives from non-engineering areas such as legal or insurance.

The conference could thus cover most of the actual questions, be it of technical or a societal relevance.

Not all authors had accepted to contribute to this edition of our accompanying book. However, the ones who did deliver excellent papers that shall give the community the whole opportunity, also the one that could not make it this time to Versailles, to learn from the information given.

Towards Sustainable, Safe, Efficient and Affordable Mobility

Remi Bastien

Abstract Car electrification, autonomous driving and connectivity are shaping the future of cars in a new mobility. Renault is facing this challenge with a complete set of electric vehicles, various concept cars and in particular initiatives in the shared mobility domain.

Keywords Sustainable · Safe · Efficient · Affordable mobility

As my eminent colleague from Mercedes, Dr. MIKULIC pointed out, mobility is not the consequence of prosperity because this is the opposite. In this sense, the automobile (and more broadly the road transport) accelerated the prosperity of the industrialized countries in the twentieth century. This acceleration finally turned into saturation with many negative effects such as accident-related mortality, environmental impacts (air quality and climate effect), invasion of urban space, congestion, the time lost in the trafic jams … Our society aspire to mobility that is more respectful of the environment and the quality of life.

Faced with these social challenges, new technologies and especially the digital transition will make possible a profound transformation of the automobile and thus of individual mobility. We are facing a triple mutation: electrification of traction, automated driving and mobility on demand.

- The first is associated with the transformation of the energy sector and thus the production of electricity. The share of renewable energies is going to grow and will cause the move of centralized production at a local production network that will integrate electric vehicles in smart grids that will require an intelligent control of this system of systems. This intelligent control will also have consequences for

R. Bastien (✉)
VP Automotive Prospective - Renault, 1 avenue du Golf, Guyancourt 78084, France
e-mail: remi.bastien@renault.com

J. Langheim (ed.), *Electronics Components and Systems for Automotive Applications*, Lecture Notes in Mobility,
https://doi.org/10.1007/978-3-030-14156-1_2

Fig. 1 A complete set of 100% electric vehicles

business models and the total cost of ownership and use of electric vehicles. Digital players will have a very wide range of applications, from modeling to invoicing, through validation, or preventive maintenance, for example (Fig. 1).

- The second, automated driving, will contribute to greatly reduce the number of accidents and enable the emergence of driverless shuttles. This theme mobilizes the automotive industry but more and more players related to digital as the GAFA.[1] It inspires countless congresses and reports: digital expertise is already heavily involved. Indeed, the rise of artificial intelligence will allow this major break in connection with a system approach of very high security and safety: it's all about controlling the risk to 10^{-9} by using, for example, formal methods. We will also have to master system of systems since the autonomous vehicles will have a high power of processing onboard associated with a connection with the infrastructures which will bring the additional data essential to ensure this very high level of safety (Fig. 2).
- Finally, the services of mobility on demand will, initially, operate connectivity and the use of big data to facilitate the sharing of vehicles. Then when driverless shuttles will mature, these services will change in depth access to mobility: the challenge is to have the flexibility of the taxi at a reduced cost by over 40%, so much more accessible and particularly to people with reduced mobility. This break can allow a complete change in the use of urban space with a drastic reduction in the number of vehicles in circulation. Here again, the digital expertise will be essential to ensure a very responsive, safe and optimal service in terms of travel (time, energy consumption), and very economical.

[1]GAFA: Google, Apple, Facebook, Amazon.

Fig. 2 Experimentation of autonomous vehicle with nuTonomy on a Renault ZOE

In addition to this change related to uses, the digital transition has a profound impact on the automotive processes. Developments and validation are increasingly using simulation, which drastically reduces the number of prototypes, compresses development time and improves robustness and quality. With CAD[2] as early as the 1980s, progress in productivity was spectacular, resulting in a halving of development time, not to mention the absorption of the increasing complexity of vehicles (crash integrity, depollution and customer benefits standards). With the three changes we need to take, the challenge of digital tools and methods is even stronger. For example, to validate the autonomous vehicle, it would be necessary to accumulate more than 15 Billion km. The simulation will be able to avoid this waste of time and energy, reducing it to confirmation rolls. Production and logistics are also impacted by «Industry 4.0» and the potential to master the processes that digital technology brings. Finally, business models are also affected with the growth of mobility-related services and the fact that the motor vehicle becomes a connected object participating in the network of IOTs.[3]

All these changes, enabled by the digital transition, will therefore revolutionize the uses where the traditional model of ownership of the automobile will be reduced to the benefit of a model of use of vehicles on demand, with the impacts on the structure of our vehicles sales. They will also profoundly transform our businesses and the structure of our jobs. Traditional skills (mechanics, stamping, foundry, electronics, etc. …) will remain essential, but we will have to master more all this know-how related to the digital and completely new for our industry. It's all about artificial intelligence, with its discipline of machine learning, cyber security through critical real-time software. The automobile industry will thus have to be open to other sectors

[2]CAD: Computer Aided Design.

[3]IOT: Internet of things.

Fig. 3 Symbioz concept car, the future of electric, autonomous and connected vehicles

of the economy and not be able to fulfill its traditional ecosystem built around the chain of suppliers.

And the key enabler of this digital transition is obviously electronics. In a congress like CESA, we can gather the best experts from automotive and electronics sector and build together the necessary conditions to make possible all these transitions. Clearly, "software is king … but there is no software without powerful hardware". The famous Moore law and the derivates as More than Moore have a bright future! (Fig. 3).

Finally, our vital challenge is to drive simultaneously all these changes to eliminate the negative effects of the automobile to offer sustainable, safe, affordable and efficient mobility. The contribution of electronics will be crucial for this transformation and thus, the automobile can continue to be a driving force for progress for our companies.

X 20 FÉVRIER 2018 • LA JAUNE ET LA ROUGE
SIMULATION ET SUPERCALCULATEURS
DOSSIER

RÉMI BASTIEN *directeur de la prospective automobile, Renault*

VERS UNE MOBILITÉ DURABLE, SÛRE, EFFICIENTE ET ABORDABLE

Les nouvelles technologies et tout spécialement la transition numérique vont rendre possible une transformation profonde de l'automobile et donc de la mobilité individuelle. Nous sommes face à une triple mutation : l'électrification de la traction, l'automatisation de la conduite et la mobilité à la demande, le tout sous-tendu par l'essor des technologies numériques.

Comme le faisait remarquer mon éminent confrère de Mercedes, Leopold Mikulic, la mobilité n'est pas la conséquence de la prospérité : c'est l'inverse. En ce sens, l'automobile (et plus largement le transport routier) a accéléré la prospérité des pays industrialisés au XXe siècle. Aujourd'hui, cette accélération s'est finalement transformée en « saturation », avec de nombreux effets négatifs comme la mortalité due aux accidents, les impacts environnementaux (qualité de l'air et effet sur le climat), l'envahissement de l'espace urbain, la congestion, le temps perdu dans les embouteillages… Nos sociétés aspirent à une mobilité plus respectueuse de l'environnement et de la qualité de vie.

« Nos sociétés aspirent à une mobilité plus respectueuse de l'environnement »

TROIS MUTATIONS FONDAMENTALES

Une première mutation concerne le secteur de l'énergie et donc la production d'électricité. La part des énergies renouvelables va croître et va entraîner le passage d'une production centralisée à un réseau de productions locales qui intégrera les véhicules électriques dans des *smart grids* qui vont nécessiter un pilotage intelligent de ce système de systèmes. Ce pilotage intelligent aura aussi des conséquences sur les modèles d'affaires et sur le coût total de possession et d'usage des véhicules électriques. Les acteurs du numérique auront un champ d'applications très vaste depuis la modélisation jusqu'à la facturation en passant par la validation, ou la maintenance préventive, par exemple.

La deuxième, l'automatisation de la conduite, va contribuer à réduire fortement le nombre d'accidents et va aussi rendre possible l'émergence de « robots-taxis ». Ce thème mobilise l'industrie automobile mais aussi de plus en plus d'acteurs liés au numérique comme les GAFA. Il suscite d'innombrables congrès et rapports : le génie numérique est d'ores et déjà fortement mis à contribution. En effet, la montée en puissance de l'intelligence artificielle va permettre cette rupture majeure en lien avec une approche sys-

REPÈRES

En France, le « véhicule à pilotage automatique » fait l'objet de l'un des 34 plans de la « nouvelle France industrielle » annoncés en octobre 2013. En 2017, un appel à projet porté par l'Ademe a été annoncé doté d'un fonds de 200 000 euros. Un programme industriel pour le véhicule connecté doit se doter d'une feuille de route avant mi-2017, portant notamment sur la cybersécurité, la gestion des données et la normalisation des infrastructures.

DR

Une gamme complète de véhicules 100 % électriques.

DR

Expérimentation de véhicule autonome avec *nuTonomy* sur une Zoé.

tème de très haute sécurité et sûreté : il s'agit de maîtriser le risque à « 10-9 » en faisant appel, par exemple, aux méthodes formelles. Nous aurons également à maîtriser des systèmes de systèmes puisque les véhicules autonomes disposeront d'une grande puissance de traitement embarquée, associée à une connexion avec les infrastructures qui apporteront les compléments de données essentielles pour assurer ce très haut niveau de sûreté.

Enfin, les services de mobilité à la demande vont, dans un premier temps, exploiter la connectivité et l'usage des *big data* pour faciliter le partage des véhicules. Ensuite, lorsque les « robots-taxis » seront à maturité, ces services vont changer en profondeur l'accès à la mobilité : l'enjeu est de disposer de la flexibilité du taxi à un coût réduit de plus 40 %, donc beaucoup plus accessible et notamment aux personnes à mobilité réduite. Cette rupture peut permettre une mutation complète de l'usage de l'espace urbain avec une réduction drastique du nombre de véhicules en circulation. Là encore, le génie numérique sera incontournable pour assurer un service très réactif, sûr, optimal en trajets (temps, consommation d'énergie), et très économique.

LA TRANSITION NUMÉRIQUE AU CŒUR DES PROCESSUS DE L'AUTOMOBILE

En plus de cette mutation liée aux usages, la transition numérique impacte en profondeur les processus de l'automobile. Les développements et la validation font appel de plus en plus à la simulation qui réduit drastiquement le nombre de prototypes, compresse les temps de développement et améliore la robustesse et la qualité. Avec la Conception assistée par ordinateur dès les années 80, les progrès en productivité ont été spectaculaires en entraînant une division par deux du temps de développement, sans compter l'absorption de la montée en complexité des véhicules (normes de choc, de dépollution et prestations client). Avec les trois mutations que nous devons conduire, l'enjeu des outils et méthodes numériques est encore plus fort.

La production et la logistique sont également impactées par l'usine 4.0 et le potentiel de maîtrise des processus qu'apporte le numérique. Enfin, les modèles d'affaires sont également touchés avec la croissance des services liés à la mobilité et au fait que le véhicule automobile devient un objet connecté participant à l'Internet des Objets.

« Assurer un service de mobilité à la demande réactif, sûr, optimal en trajets et économique »

Toutes ces mutations, permises par la transition numérique, vont donc révolutionner les usages où le modèle traditionnel de possession de l'automobile se réduira au profit d'un modèle d'usage des véhicules à la demande, avec les impacts sur la structure de nos ventes. Elles vont également transformer en profondeur nos métiers et la structure de nos emplois. Les compétences traditionnelles (mécanique, emboutissage, fonderie, électronique, etc.) resteront indispensables, mais nous allons devoir maîtriser en plus tout ce savoir-faire lié au numérique et complètement nouveau pour notre industrie. Cela va de l'intelligence artificielle, avec sa discipline du *deep learning*, à la cybersécurité en passant par les logiciels critiques temps réel. L'industrie automobile va devoir ainsi s'ouvrir aux autres secteurs de l'économie et ne pourra plus se satisfaire de son écosystème traditionnel constitué autour de la chaîne de ses fournisseurs.

Notre enjeu vital est donc de conduire simultanément toutes ces mutations afin d'éliminer les effets négatifs de l'automobile pour faire apparaître une mobilité durable, sûre, abordable et efficiente. L'apport du numérique sera déterminant pour cette transformation et ainsi, l'automobile pourra continuer à être un moteur de progrès pour nos sociétés. ■

LE NUMÉRIQUE AU SERVICE DE LA VALIDATION

Pour valider le véhicule autonome, il faudrait accumuler plus de 15 milliards de kilomètres. La simulation pourra éviter cette dépense de temps et d'énergie, en la réduisant à des roulages de confirmation.

DR

Symbioz concept, le futur du véhicule électrique, autonome et connecté.

Contribution of Light and Heavy Vehicles to Reduction of Energy Demand and CO_2 Emissions by 2035 in the World

Jean-Luc Brossard and Gabriel Duquesnoy

Abstract The French Automotive Organization PFA, in order to appreciate the evolution of CO_2 emissions in the world by 2035, have built with the BDO-BIPE, a projection model of the parks, sales and technological mix of light vehicles and trucks. The results of the model, is produced annually in order to integrate recent changes in the different markets and to take into account the most up-to-date and plausible regulatory and technological developments over the time horizon.

The main conclusions of the 2018 study are as follows:

- The inflection of the road transport CO_2 curve of the global road transport sector is allowed with our green scenarios, on the one hand by the progress of the electrification of light vehicle powertrains (VL) and on the other hand, by the expected slowdown in growth dynamics of the VL fleet in China. These two combined effects make possible, after 2023, to offset the effect of the significant increase in the number of Light Vehicles contribution and of the Heavy Vehicle fleet worldwide.
- Europe and North America account for 90% of CO_2 emission reductions. China, Africa and Asia excluding the OECD (including notably India) account for more than 90% of the increases in emissions.
- Electrified technologies will be the main contributors to the reduction of CO_2 emissions over the 2020–2035 period. These technologies have sufficiently high market shares only in scenarios where incentive levers are maintained or put in place over this period.
- The electrification of the automotive sector, as well as the development of other alternative energy sources—biofuel, natural gas and hydrogen (in the context of carbon-free energy production)—the development of new mobility offers (car-sharing, ride-sharing), access restriction measures, see the progressive ban on the sale of 100% thermal vehicles (eg in Europe), are in 2035 much more effective to bend the curve of CO_2 emissions than the tightening of standards emissions.

J.-L. Brossard (✉)
PFA, 2 Rue de Presbourg, 75008 PARIS, France
e-mail: jeanluc.brossard@pfa-auto.fr

G. Duquesnoy
BDO BIPE Advisory, 43-47 Ave de la Grande-Armée, 750116 Paris, France

J. Langheim (ed.), *Electronics Components and Systems for Automotive Applications*, Lecture Notes in Mobility,
https://doi.org/10.1007/978-3-030-14156-1_3

- For the nominal scenario, the total share of electro mobility in light vehicle sales in 2035 is estimated at 20% worldwide. The proportion of BEVs represents 12% of sales worldwide (the share of PHEVs being at 8%). In Europe, electro mobility will account 35% sales in 2035 (including 25% BEV and 10%PHEV).

Keywords Mix · Energy · Powertrain · Baterries electric vehicles · Plug in hybrid · $C0_2$

1 Introduction

To assess changes in CO_2 emissions in the world by 2035 a model projecting running fleets volumes, sales and technological mixes of light vehicles and heavy vehicles has been created. This technical note, which presents the results of the model, is produced each year to incorporate recent changes in the various markets, and to take into account those changes in terms of regulations and technology which are most up-to-date and most plausible for the time frame under analysis.

The 2018 version of the model's major hypotheses now includes stoppage, after 2030, of sales of 100% diesel and petrol internal combustion engine (ICE) vehicles in Europe, the future appearance of a broader range of light vehicles powered by natural gas in this geographical area or, alternatively, the incorporation of regulations on a country-by-country basis [and in particular regulations specific to light commercial vehicles (LCV)].

The main conclusions drawn from the 2018 analysis are as follows:

- A downward trend of CO_2 emissions in the road transport sector at world level is permitted under green scenarios, firstly due to the increase of electrification of light vehicles (LV) and secondly due the expected slow-down of the patterns of growth of LV vehicles in China. After 2023 these two effects combined enable the effect of the substantial increase of volumes of LV and Heavy Vehicles (HV) at world level to be compensated.
- Europe and North America account for 90% of reductions of CO_2 emissions. China, Africa and Asia outside the OECD (including India, in particular), for their part, account for more than 90% of rises in emissions.
- Electric technologies will be the main contributors to reduction of CO_2 emissions over the period 2020–2035. These technologies have sufficiently high market shares only in scenarios in which incentive mechanisms are maintained or established over this period.
- In 2035 the electrification of the motor vehicle sector, and the growth of other alternative sources of energy—biofuel, natural gas and hydrogen (in connection with zero-carbon energy production), the growth of new mobility solutions (car-sharing and ride-sharing), measures to restrict access, or gradual prohibition of sales of 100% internal combustion engine vehicles (e.g. in Europe) are much more effective at reducing CO_2 emissions than toughening of emissions standards.

- In 2030 and 2035 the total share of electro-mobility in sales of light vehicles is estimated at, respectively, 17 and 20% across the world. The proportion of BEV represents in 2030 and 2035 10 and 12% of world sales (the proportion of PHEV is 7 and 8%). In Europe electromobility will represent 29% of sales in 2030 (20% of which BEV and 9% PHEV), and 35% of sales in 2035 (25% of which BEV and 10% PHEV).

2 Context and Motivations

Following on from the decisions of COP21, most States throughout the world have committed themselves to limiting their greenhouse gas emissions to fight climate change. In 2015 road transport (private vehicles, light commercial vehicles and heavy vehicles) are responsible for 18%[1] of CO_2 emissions, compared to 15% in 1990. This sector is therefore called upon to attain the goals defined in Paris at COP21. At European level the energy-climate package has set ambitious goals for Europe which have been used as the basis for its international commitments. In the transport field the EU has set itself the goal of 10% of energy of renewable origin in 2020, and could determine on a goal of 14% in 2030.[2]

The energy efficiency of sold vehicles has already substantially improved (26% of efficiency gains on average in Europe per new vehicle between 2005 and 2016[3]), due to improvements with combustion engines, to the growth and sale of new powertrains (hybrid systems and electric vehicles) and improvements with vehicles (mass, aerodynamics and rolling resistance). But, simultaneously, motor vehicle volumes and usage continue to grow, propelled by the growth of a middle class in still-emerging markets (China, South America, ASEAN, etc.), and by the ever-increasing attraction of motor vehicle models of the SUV type. Faced with increasingly demanding regulations, the Motor Vehicle Industry must therefore come up with new efficiency gains by 2035.

In 2015 and 2016 the French motor vehicle sector initiated a working programme **to assess the contribution of road transport to reduction in energy demand and CO_2 emissions across the world**. The analysis carried out has many goals:

- Producing a panorama of energy demand for each area and type of vehicle by 2035, with an analysis of risks and opportunities in relation to the variables
- Developing a reasoned and independent view of the energy future of road transport and its impact in terms of CO_2 emissions
- Pooling and coordinating the research efforts of industrial companies for future technologies in the power chain field.

---

[1] Source: IEA, CO_2 Emissions from Fuel Combustion Highlights 2017. CO_2 emissions from road transport represent 5792 Mt CO_2 in 2013, with total emissions 32,294 Mt CO_2.

[2] GAIN USDA 2018, EU Biofuels Annual 2018.

[3] Source: ICCT, CO_2 emissions from new passenger cars in the EU: Car manufacturers' performance in 2016, July 2017. Average consumption rose from 6.9 l/100 km in 2005 to 5.1 l/100 km in 2016.

To undertake this study the sector contacted BIPE, a research and consultancy firm, in order to devise a rational and independent view of changing motor vehicle volumes, the speed at which vehicles are renewed, and segmentation and distribution of them for each type of energy and powertrain. Similarly, changes to the production mix of the various energies were analysed for each area and for certain countries. It was thus possible to create a model to project total energy demand; total CO_2 emissions were deduced from it.

3 Methodology and Scenarios

The model takes account of 15 geographical areas comprising all the world markets and 46 market segments (7 for PC, 4 for LCV, 35 for HV[4]). 15 powertrain are analysed: petrol and diesel combustion engines (including start/stops and micro-hybrids), all combinations of hybridation (low-voltage mild, 12 and 48 V, full, plug-in, range-extender), zero-emission vehicles (electric and fuel cell), gas-powered vehicles (NGV, LPG). The study relates to the period 2017–2035 for all segments of vehicles studied (light and heavy).

The general principle of the methodology is explained in below (Fig. 1).

The motorisation rate (number of vehicles per 1000 inhabitants) as a function of GDP per inhabitant is an S-shaped curve which can be configured by a rate of saturation, a take-off GDP, and a take-off speed, which are specific for each area (see Fig. 2). The ownership rate is very low for low GDPs per inhabitant; then from a

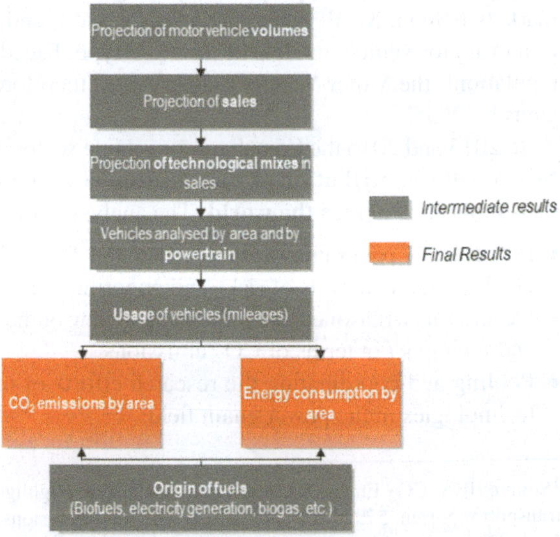

Fig. 1 General outline of the projection approach

[4]See Appendix 1 for the definition of vehicle segments.

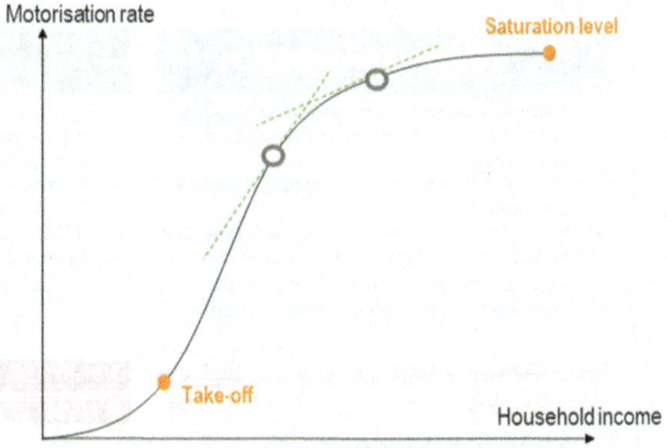

Fig. 2 S-shaped curve for calculating volumes

threshold it starts to climb, and after that changes twice as fast as GDP. Its rise slows, reaching a threshold, and the rate will no longer change at the same speed as GDP, or may even decline. Due to growth hypotheses it is therefore possible to deduce the motorisation rate for each area and scenario, and therefore motor vehicle numbers by 2035. The scale of registrations and the structure of the volumes by age are then determined using a scrappage law.

After precise and exhaustive inventories of the costs for purchase and for usage of the various powertrains (TCO—Total Cost of Ownership), BIPE's models rely on formalising client purchase trade-offs. The models are predictive and are fed with explanatory variables (technological, fiscal, macro-economic, mobility habits, etc.) to project sales and volumes which are structured according to powertrain. From the projection concerning volumes, combined with usages (mileages) and unit consumptions, a demand for fuel and CO_2 emissions is obtained.

To assess the impacts of the uncertainties relating to the context (macro-economic changes, regulations, etc.), 4 prospective scenarios were devised. Each of the scenarios makes reference to differentiated and consistent hypotheses concerning possible changes of the global environment and underlying factors of motor vehicle markets (cf. Fig. 3).

The results of technological mix and of total energy demand were projected in each of these scenarios, which will enable the most effective medium- and long-term measures to limit and then reduce CO_2 emissions to be determined. The reference scenario is the **Green Constraint** scenario, with low economic growth and a high level of environmental regulation. Indeed, since the economic crisis world growth has stayed fragile (approximately 3% per year, compared to 4–5% before the crisis, with a substantial slow-down of the BRICS), and at the same time environmental conferences such as COP21 are succeeding in bringing together most of the world powers and in causing them to reach a consensus.

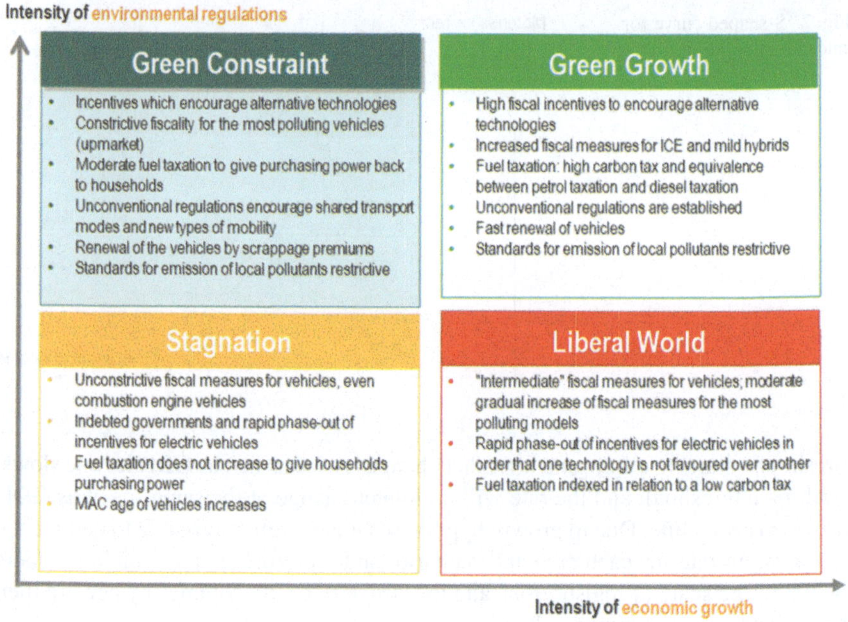

Fig. 3 Projection scenarios

4 Results

4.1 Demand for Mobility

Demand for mobility is an essential result of the model (cf. Figs. 4 and 5). **In mature motor vehicle markets total numbers have reached a level of saturation** (500 LV per 1000 inhabitants in Europe, 800 in the United States), which depends on the specific features of each country (e.g. urban organisation, public transport, geography), and changes to motor vehicle volumes and usages (unit mileages) are now no longer correlated with economic growth in these areas. In addition, demands for mobility (persons and goods) are subject to drastic optimisation against a background high energy price volatility.

Conversely, **in the Emerging countries** (e.g. China, India, Brazil and Russia) a **high degree of dependency is observed between economic growth and growth of total demand for mobility**. Initially, economic growth is accompanied by a sudden rise in the numbers of utility vehicles (light and heavy) which enable the industrial fabric and the economy in general to grow. The growth in incomes is then accompanied by the emergence of a middle class which purchases private vehicles.

The scenarios are therefore highly differentiated (cf. Fig. 6): in scenarios where there is strong growth the total LV (Light Vehicles) motor vehicle number is close to 2.1 billion vehicles (compared to 1.2 billion currently), with markets of over **150**

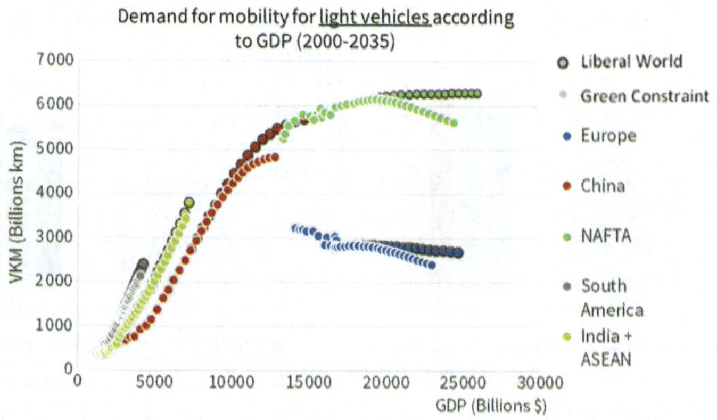

Fig. 4 Demand for mobility for LV for each area in the ***Green Constraint*** and ***Liberal World*** scenarios

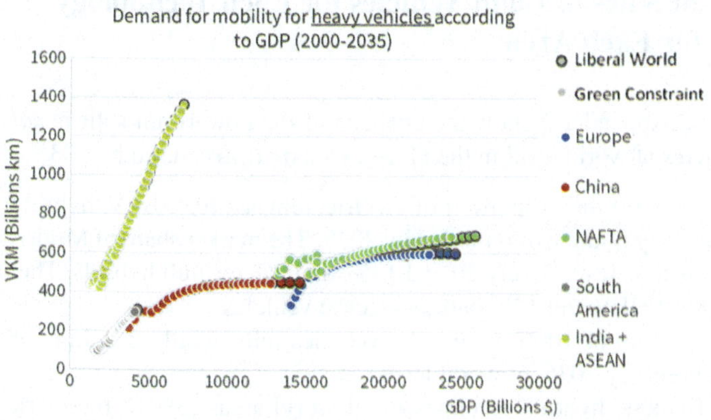

Fig. 5 Demand for mobility for HV for each area in the ***Green Constraint*** and ***Liberal World*** scenarios

million sales of new vehicles per year in 2035 (compared to 95 million in 2017); in the Stagnation and **Green Constraint** scenarios total world numbers are close to 1.9 billion vehicles, with **a market of close to 130 million in 2035**. In these scenarios world growth is highly subdued and total demand for mobility declines as a result. The impacts of new forms of mobility (e.g. car-sharing and ride-sharing) are included (reduction of motor vehicle numbers by approximately 10% by 2035), with a reduction of total numbers in the mature areas (Europe, Japan), and stable in the United States. These emission reductions are even more substantial if governmental regulations are stronger, as in the green scenarios.

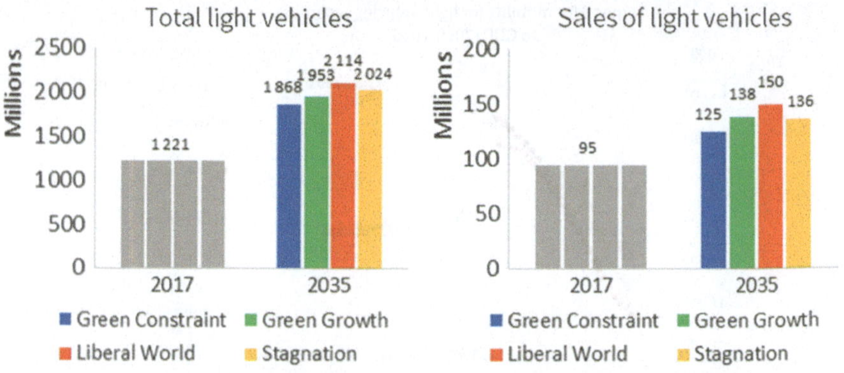

Fig. 6 Total number and annual sales of LV (PC + LCV) in 2035

5 Mix of Sales of Light Vehicles for Each Technology and for Each Area

Table 1 and Appendix 2 show the changes of the powertrain split in ***sales*** of ***new light vehicles*** at world level in the **Green Constraint**scenario by 2035.

- In this scenario the proportion of electric vehicles (e.g. BEV, hybrids) accounts for a majority of the world market by 2035. The market share of Mild 12 and 48 V hybrids is thus respectively 20 and 13%, and 2% for Full hybrids. There are even nearly 8% PHEV and 12% battery electric vehicles.
- The proportion of 100% pure ICEs vehicles falls greatly, reaching 38%, 3% of which diesel, by 2035, at world level.
- Diesel (include hybrids) declines in all world areas (5% of the market in 2035 compared to 17% in 2017, and nearly 20% before "Diesel Gate"): very strict pollution reduction standards and regulations in urban areas which are penalising residual values in Europe, deregulation of prices of fuels in India, and growth of small segments in ASEAN countries.
- Gas grows significantly in Europe, increasing from less than 1% to over 10% between 2017 and 2035. However, this market share is subject to the hypothesis that 100% ICEs vehicles will cease being sold in Europe, which would not affect CNG vehicles. Over the same period it grows more moderately in certain growth markets, particularly in India, rising from 5% to nearly 10%, and at world level CNG will increase from 1 to 4%.
- As for electrification, low-hybridation engines (mild hybrids) are ahead of the pack, and can be seen as the natural development for start/stop systems. They account for 32% of the world market in 2035. Full hybrids are stable and will have 2% of the market in 18 years, due in particular to the American and Japanese markets.
- Finally, highly electrified vehicles (BEV and PHEV) are growing in areas which impose severe regulation of CO_2 emissions for new vehicles (Europe, USA, China

Table 1 Mix of power trains of LV sales (**PC + LCV**), world, in the *Green Constraint* scenario

Technology	World		China		Europe		NAFTA		India	
	2017 (%)	2035 (%)	2017 (%)	2035 (%)	2017 (%)	2035 (%)	2017 (%)	2035 (%)	2017 (%)	2035 (%)
Gasoline 100% combustion engine	76.3	35.8	87.4	29.9	45.6	0.0	92.0	25.6	51.0	50.7
Electric gasoline[a]	3.4	41.4	3.0	47.0	3.0	57.2	4.4	60.3	0.0	24.1
Diesel 100% combustion engine	17.2	2.5	8.0	0.8	49.8	0.0	1.5	2.3	43.1	3.2
Electric diesel[a]	0.3	2.6	0.4	2.2	0.1	5.4	0.2	3.4	0.0	2.4
CNG	1.2	4.0	0.0	0.0	0.5	11.7	0.6	0.5	5.2	8.7
LPG	0.9	1.7	0.0	0.0	0.6	1.1	0.5	0.2	0.5	1.9
BEV	0.6	11.9	1.2	20.1	0.5	24.4	0.8	7.6	0.0	9.1
Hydrogen	0.0	0.1	0.0	0.0	0.0	0.2	0.0	0.1	0.0	0.1

[a]Including low-hybridation technologies (mild 12 V and 48 V hybrids) and full hybrids and rechargeable hybrids. The results are shown in detail in Appendix 2

and Japan). **Electromobility (BEV and PHEV) will account for 20% of sales by 2035 in the reference scenario.** In total, PHEV take 8% of the market, supported by the United States and Europe, while BEV become well-established in China (20%) and Europe (24%). The latter benefit in particular from changes in regulations (e.g. policies to remove 100% ICEs vehicles in Europe by around 2030, and toughening of CO_2 emissions standards). The proportion of BEV at world level reaches 12% in 2035. VE will still be limited by a number of restraints, in particular the perception by consumers of vehicle recharging times in certain geographical areas; however, these should gradually dissipate by 2035 due to positive developments in terms of autonomy, charging times and the cost of batteries.

- In Europe the anticipation of the toughening of CAFE regulations after 2025 is leading to a gradual removal of 100% petrol and diesel combustion engine vehicles, and from 2020 town centre access restrictions will be put in place for the most polluting vehicles. The effects of these two mechanisms are an acceleration of declines of market shares of diesel for the area, and increased interest in gas technologies for LV segments which could satisfy consumers who are most sensitive to purchase prices, and whose usage patterns are less compatible with BEV.
- The results of the market shares of CNG, which are 11.7% in 2035 in Europe, are obtained by considering that this technology would not be affected by future restrictions for combustion engine vehicles.

6 Energy and CO_2 Emissions Efficiency: "from the Well to the Wheel"[5]

In the Stagnation and Liberal World scenarios world energy demand in 2035 increases respectively by 13 and 17% (basis 2017, cf. Fig. 7). In the Green Growth and Green Constraint scenarios demand declines by 3%. In terms of CO_2 emissions relating to energy use by road transport (cf. Fig. 8, excluding figures for vehicle production), these increase by over 10% in the Stagnation and Liberal World scenarios, stabilise in the Green Constraint scenario and decline by 3% in the Green Growth scenario. **The introduction of electric vehicles and the growth of gas-powered vehicles enable the emissions curve to be levelled off in the green scenarios, but do not compensate sufficiently for the large increase in world vehicle numbers in the other two scenarios.** This reflects 2 opposing effects: high growth of motor vehicle numbers in Emerging countries, compensated partially by efforts made in relation to efficiency of combustion engine vehicles and the introduction of low-emission vehicles. However, this type of progress takes time to become widely adopted, and to modify consumption, given that a given pool of vehicles is renewed only every 17 years on average. It is therefore to be expected that the substantial efforts made by the manufacturers from 2000 onwards should only bear fruit several years later.

[5]Vehicle production is not included when rating efficiency.

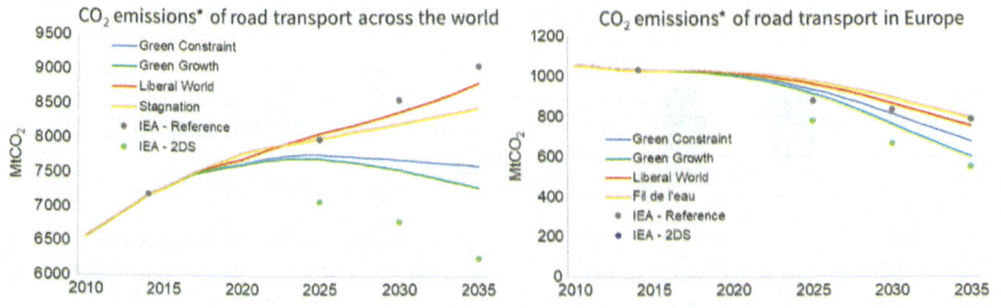

Fig. 7 CO_2 emissions of road transport throughout the world and in Europe (*: excluding figures for vehicle production)

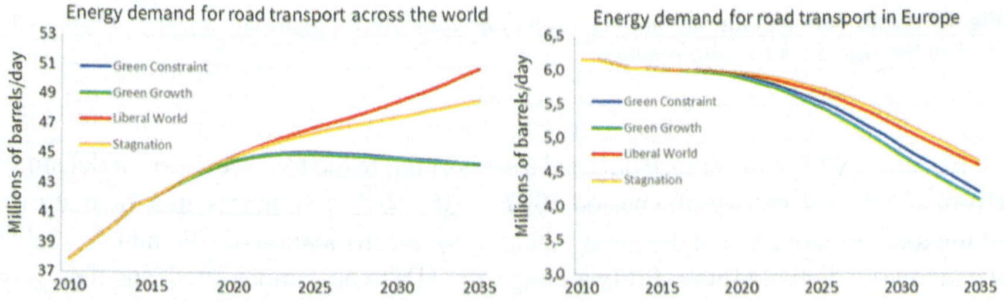

Fig. 8 Road transport energy demand in the world and in Europe

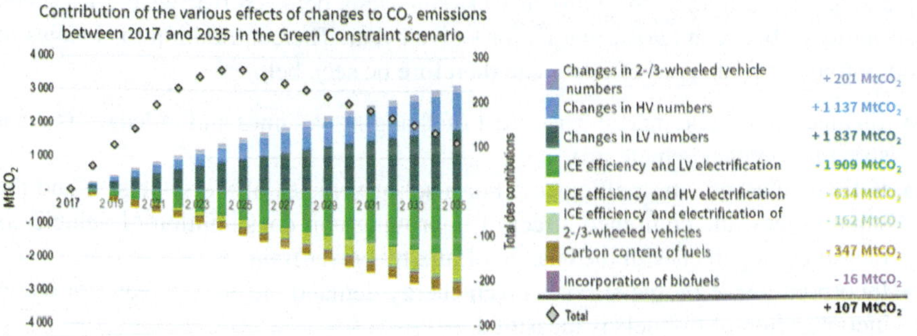

Fig. 9 Contribution of the various effects on CO_2 emissions relating to road transport between 2017 and 2035 in the *Green Constraint*scenario

The finding for world CO_2 emissions is differentiated by category of vehicles (cf. Fig. 9):

- Rise in emissions for heavy vehicles (17–19% growth in 2035 compared to 2017), due to an increasing demand for mobility,
- In the "green" scenarios emissions for light vehicles (LV) decline (between 8% for Green Constraint and 13% for Green Growth). LV emissions increase by 7–15% respectively in the Stagnation and Liberal World scenarios.

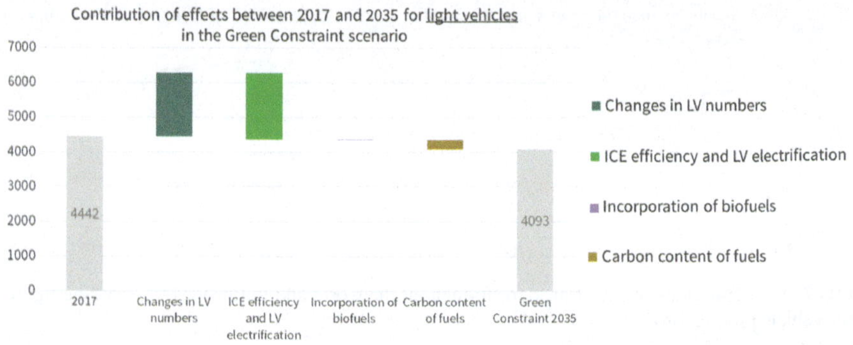

Fig. 10 Contribution of the various effects on CO_2 emissions for light vehicles between 2017 and 2035 in the ***Green Constraint***scenario

In Europe CO_2 emissions from road transport fall in the four scenarios under study (from −23 to −40% over the period 2017–2035, cf. Fig. 8). In this area the maturity of the total pool of LV and the great potential for electrification of LV and HV enable the increased demand for mobility arising from HV to be compensated, and therefore a more than significant improvement of road transport sector emissions.

The differentiation of the effects presented in the following figures allows comparisons of the figures for gains of CO_2 emissions between two years of a given scenario, or between two scenarios for a given year. The contribution to changes of emissions for four types of effect can therefore be seen below:

- Changes in vehicle mixes: The effect of changing volumes and mileages is measured for a given vehicle mix.
- ICE and electrification efficiency: For a set of vehicles of the same size and the same usages, the combined effect of improving unit consumption of vehicles is measured, together with the change of mix by power train.
- Incorporation of biofuels: For a given energy demand the effect greater or lesser incorporation of biofuels is measured.
- Carbon intensity effect: For a given fuel consumption the effect of changing their carbon content is measured (e.g. zero emission of the energy mix relating to electricity generation).

Biofuels (1G + 2G) account for 10% of world fuel consumption in energy terms in 2035, compared to 6% in 2017 (cf. Appendix 10).

Thus, despite the strong emergence of a middle class in certain countries, which is tending to increase vehicle numbers, the efforts made by the entire value chain for LV bears fruit for both "green" scenarios. The significant improvements in combustion engine consumption, coupled with changes to the powertrain mix overall, counterbalance the growth of light vehicle numbers (cf. Fig. 10), which is not the case for the Liberal World and Stagnation scenarios.

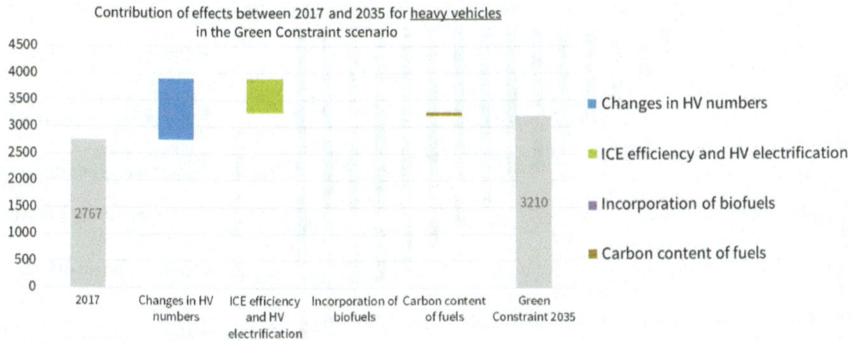

Fig. 11 Contribution of the various effects on CO_2 emissions for heavy vehicles between 2017 and 2035 in the ***Green Constraint*** scenario

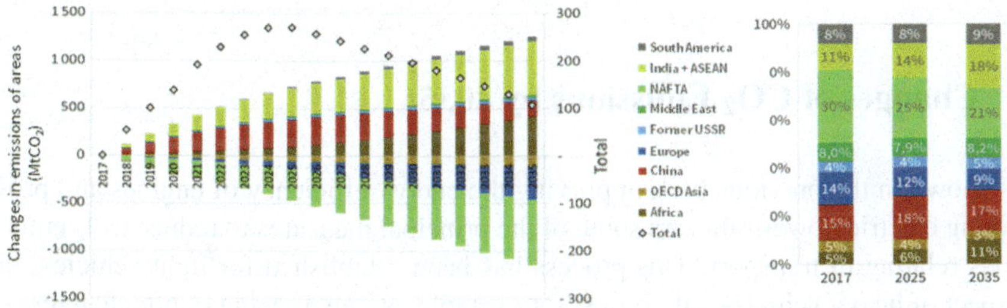

Fig. 12 Changes in CO_2 emissions relating to road transport for each area between 2017 and 2035, and proportion of the area in CO_2 emissions in the ***Green Constraint*** scenario

Thus, in OECD areas, where vehicle numbers are relatively stable, **electrification of vehicles and improved efficiency of combustion engines are the main mechanisms for reducing emissions**.

This is not the case with heavy vehicles, where the less significant progress made with efficiency of new vehicles does not enable to high rise in vehicle numbers to be compensated (cf. Fig. 11). Indeed, unlike LV, where technological progress enables to emissions to be reduced despite the growth in numbers, gains made with HV enable only some 56% of the growth of emissions relating to the increased number of HV in use to be absorbed, despite a strong emergence of electric in certain HV segments such as buses.

In Developing and Emerging countries the increased vehicle numbers are not compensated, but at world level the major efforts made in Europe and North America enable the global emissions curve to level off, as the patterns of growth of LV numbers are weakening in China (cf. Fig. 12) in both "green" scenarios. Emission reductions are to be found in the OECD countries, which now account for 49% of total emissions. The increase in numbers allowed by economic growth in Emerging countries explains the increase in emissions between 2017 and 2035. The latter will be the sources of nearly 60% of global CO_2 emissions in 2035.

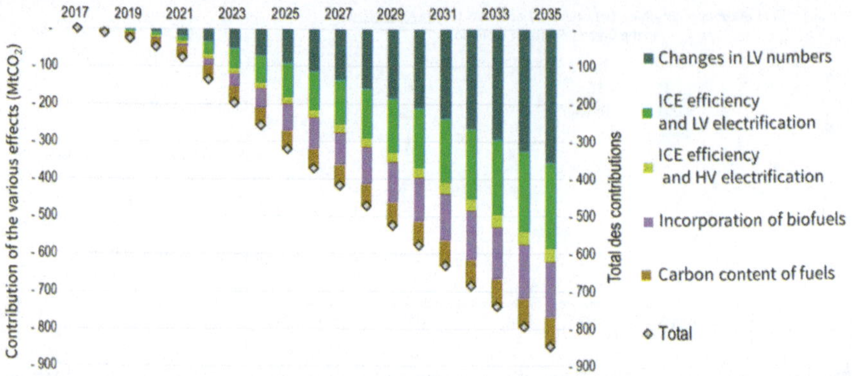

Fig. 13 Contribution of the various effects to reduction of CO_2 emissions relating to road transport between 2017 and 2035 in the ***Green Constraint*** and *Stagnation*scenarios

7 Changes of CO_2 Emissions by 2035

As shown in the previous part, improving the energy efficiency of engines and promoting electric powertrains are some of the principal measures to reduce CO_2 emissions relating to transport. This process has been established for light vehicles. In respect of heavy vehicles, city buses, for example, are also seeing major changes of the power train mix in sales, which are tending towards all-electric across the world.

The effects of policies established in the medium term to reduce CO_2 emissions can also be measured by comparing two scenarios with equal growth, one "green" and one not "green". In both scenarios the impact of the manufacturers' efforts to reduce consumption of combustion engines is the same, and only the rate of electrification changes the results substantially over the period 2017–2035.

In 2035 reduced carbon intensity of alternative energy sources (biofuel, electricity, biomethane and hydrogen), the development of new types of mobility (e.g. car-sharing and ride-sharing), and access restriction measures are much more effective in lowering the emissions curve than toughening of CO_2 emissions standards (cf. Fig. 13).

32% of the difference in emissions of light vehicles between the Stagnation scenario and the Green Constraint scenario, i.e. a difference of 229 tonnes, derives from differences in penetration of more efficient (low hybridation) and highly electrified vehicles (cf. Fig. 14).

Electric technologies will be the main contributors to reduction of CO_2 emissions over the period 2030–2040. These technologies have sufficiently high market shares only in scenarios in which incentive mechanisms are maintained or established between 2020 and 2030. In addition, the establishment of very severe CAFE standards after 2025 would make the attainment of these targets highly dependent on the short- and medium-term development of a recharging infrastructure which is capable of getting the market to accept vehicles with very low emissions rapidly,

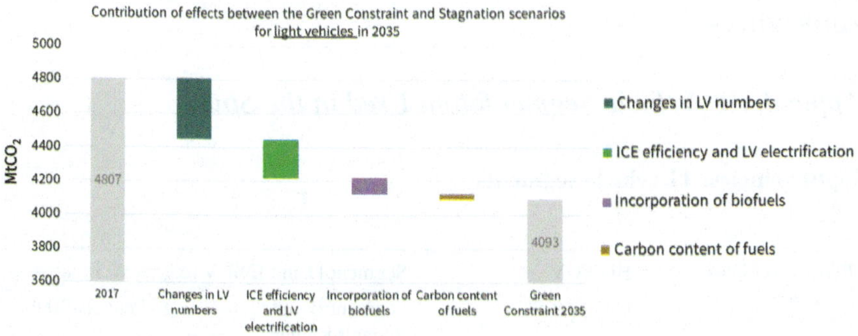

Fig. 14 Comparison of CO_2 emissions of road transport of the *Green Constraint* and *Stagnation* scenarios in 2035, by type of effect

and on a large scale. This would also presuppose significant breakthroughs in terms of research and development relating to battery technologies, to make electrification economically accessible.

8 Conclusions of the Study

1. A downward trend of CO_2 emissions in the road transport sector at world level can be achieved, firstly, due to increased electrification of power chains of light vehicles (LV), and secondly due the expected slow-down of the patterns of growth of LV vehicles in China. After 2023 these two effects combined will enable the effect of the substantial increase of volumes of LV and Heavy Vehicles (HV) at world level to be compensated.
2. Over the period 2017–2035 Europe and North America are responsible for 90% of the reduction of CO_2 emissions, and Africa and Asia for most of the increase.
3. Electric technologies will be the main contributors to reduction of CO_2 emissions after 2025. These technologies have sufficiently high market shares only in scenarios in which incentive mechanisms are maintained or established between 2020 and 2035.
4. In 2035 reduced carbon intensity of alternative energy sources (biofuel, electricity, biomethane and hydrogen), the development of new types of mobility (e.g. car-sharing and ride-sharing), access restriction measures and the gradual cessation of sales of 100% combustion engine vehicles in Europe are much more effective in lowering the emissions curve than toughening of CO_2 emissions standards.
5. In 2030 and 2035 the proportion of electromobility of light vehicles is estimated, respectively, at 17 and 20% of BEV and PHEV at world level (including 7 and 8% of PHEV) and at 29 and 35% in Europe (including 9 and 10% of PHEV in 2030 and 2035).

Appendices

Appendix 1: Vehicle Segmentation Used in the Study

Light vehicles: 11 vehicle segments

Private vehicles *7 segments*	Big SUV	Segment D and E SUV (e.g.: VW Touareg)
	Compact SUV	Segment B and C SUV (e.g.: Peugeot 2008, Renault Kadjar)
	Luxury car	Upmarket vehicles (e.g.: BMW Series-5, Audi A8)
	Large car	Segment D and E saloons or MPVs (e.g.: Renault Espace, Peugeot 508)
	Medium car	Segment C saloons or MPVs (e.g.: Renault Espace, Peugeot 308)
	Small car	Segment B saloons or MPVs (e.g.: Renault Clio, Peugeot 208)
	Extra small car	Segment A saloons or MPVs (e.g.: Renault Twingo, Peugeot 108)
Light commercial vehicles *4 segments*	Minivans	"Low-cost" LCV developed for Emerging markets (e.g.: Wuling Hongguang)
	Commercial vehicle	Segment C LCV (e.g.: Renault Kangoo, Peugeot Partner)
	Medium van	Vans with mass of less than 3.5 tonnes (e.g.: Renault Traffic, Peugeot Boxer)
	Heavy van	Vans with mass of between 3.5 tonnes and 5 tonnes (e.g.: Renault Master, Peugeot Boxer)

Heavy vehicles: 30 vehicle segments

6 usages: Construction, Long-Distance Freight, Regional Deliveries, Urban Deliveries, Buses, Coaches.

×

5 Tonnage/Power Categories: <25 kW, <225 kW, <325 kW, <425 kW, >425 kW

Appendix 2: Distribution of Sales of New Light Vehicles, by Power Train Mode, in the Green Constraint Scenario

LV	*World*					Europe				
%	*2017*	*2020*	*2025*	*2030*	*2035*	2017	2020	2025	2030	2035
Gasoline	*76.3*	*65.2*	*44.7*	*35.7*	*35.8*	45.6	47.0	35.6	3.0	0.0
Gasoline mild (12 V)	*0.5*	*8.7*	*19.5*	*21.9*	*20.1*	0.0	6.0	17.7	31.5	28.8
Gasoline mild (48 V)	*0.2*	*4.8*	*11.1*	*12.6*	*11.5*	0.0	2.9	9.6	17.9	17.2
Gasoline Full	*2.1*	*1.5*	*1.8*	*1.9*	*2.0*	2.3	1.2	1.0	1.2	1.2
Gasoline PHEV	*0.4*	*1.8*	*4.0*	*5.8*	*6.7*	0.6	2.6	4.4	8.0	9.2
Gasoline EREV	*0.2*	*0.4*	*0.7*	*0.9*	*1.1*	0.1	0.3	0.5	0.7	0.8
Diesel	*17.2*	*11.5*	*4.9*	*2.9*	*2.5*	49.8	32.9	11.6	0.7	0.0

(continued)

(continued)

LV	*World*					Europe				
%	*2017*	*2020*	*2025*	*2030*	*2035*	2017	2020	2025	2030	2035
Diesel mild (48 V)	*0.1*	*0.9*	*1.8*	*2.0*	*1.6*	0.0	2.0	4.1	5.4	4.4
Diesel PHEV	*0.2*	*0.5*	*0.9*	*1.1*	*1.0*	0.1	0.4	0.7	1.1	1.0
CNG	*1.2*	*1.5*	*2.3*	*3.4*	*4.0*	0.5	0.9	4.3	9.4	11.7
LPG	*0.9*	*1.0*	*1.2*	*1.5*	*1.7*	0.6	0.9	1.1	1.2	1.1
EV low range	*0.6*	*1.9*	*5.3*	*7.8*	*9.1*	0.5	2.6	8.2	16.6	19.5
EV High range	*0.0*	*0.3*	*1.7*	*2.4*	*2.8*	0.0	0.2	1.3	3.5	4.9
Fuel cell	*0.0*	*0.0*	*0.0*	*0.0*	*0.1*	0.0	0.0	0.0	0.0	0.2
LV	China					USA				
%	2017	2020	2025	2030	2035	2017	2020	2025	2030	2035
Gasoline	87.4	71.0	42.1	32.4	29.9	91.4	69.4	39.4	27.0	22.8
Gasoline mild (12 V)	1.8	10.7	21.8	22.8	20.9	0.0	10.3	24.3	24.1	20.8
Gasoline mild (48 V)	0.8	5.4	12.7	14.9	14.8	0.0	8.3	20.5	21.6	19.4
Gasoline Full	0.0	0.4	0.8	1.0	1.1	3.2	2.7	3.9	4.9	5.0
Gasoline PHEV	0.2	1.6	5.1	6.7	8.1	1.0	3.0	4.9	11.1	16.0
Gasoline EREV	0.2	0.4	1.3	1.9	2.1	0.5	0.9	0.8	1.6	2.5
Diesel	8.0	6.0	1.7	1.0	0.8	1.6	1.2	1.4	1.6	1.4
Diesel mild (48 V)	0.2	0.8	1.1	0.9	0.8	0.0	0.2	1.0	1.5	1.4
Diesel PHEV	0.2	0.7	1.3	1.7	1.4	0.2	0.4	0.7	1.3	1.7
CNG	0.0	0.0	0.0	0.0	0.0	0.7	0.5	0.4	0.4	0.4
LPG	0.0	0.0	0.0	0.0	0.0	0.5	0.3	0.2	0.2	0.1
EV low range	1.2	2.1	8.1	12.1	15.1	0.9	2.7	2.2	3.5	5.9
EV High range	0.0	0.8	4.0	4.6	5.0	0.0	0.1	0.5	1.2	2.4
Fuel cell	0.0	0.0	0.0	0.0	0.0	0.0	0.0	0.0	0.0	0.1
LV	Japan					India				
%	2017	2020	2025	2030	2035	2017	2020	2025	2030	2035
Gasoline	65.1	57.6	32.9	23.5	19.8	51.0	56.8	50.7	51.0	50.7
Gasoline mild (12 V)	0.0	6.6	15.8	19.7	16.2	0.0	6.3	18.2	18.8	18.7
Gasoline mild (48 V)	0.0	4.5	13.3	13.7	11.7	0.0	0.8	2.3	2.7	2.6
Gasoline full	18.9	11.0	6.3	4.4	4.2	0.0	0.0	0.1	0.1	0.1
Gasoline PHEV	0.6	1.8	4.1	5.0	5.8	0.0	1.0	3.4	3.1	2.7

(continued)

(continued)

LV	Japan					India				
%	2017	2020	2025	2030	2035	2017	2020	2025	2030	2035
Gasoline EREV	0.1	0.3	0.8	0.8	0.8	0.0	0.0	0.0	0.0	0.0
Diesel	13.9	10.3	5.5	4.4	3.6	43.1	23.3	7.3	4.4	3.2
Diesel mild (48 V)	0.0	1.0	3.2	3.8	3.3	0.0	0.8	1.5	1.5	1.1
Diesel PHEV	0.2	0.4	0.7	1.0	1.0	0.0	0.6	1.6	1.5	1.3
CNG	0.5	1.2	1.6	2.3	1.7	5.2	7.6	7.3	8.1	8.7
LPG	0.6	0.9	1.1	1.2	1.2	0.5	0.8	1.0	1.4	1.9
EV low range	0.2	3.8	12.1	15.0	20.8	0.0	1.9	5.7	6.0	7.3
EV high range	0.0	0.5	2.6	5.1	9.1	0.0	0.1	0.9	1.4	1.8
fuel cell	0.0	0.2	0.0	0.1	0.7	0.0	0.0	0.0	0.0	0.1
LV	Brazil					Russia				
%	2017	2020	2025	2030	2035	2017	2020	2025	2030	2035
Gasoline	96.2	82.5	72.9	67.6	68.5	85.6	67.6	44.6	40.4	42.2
Gasoline mild (12 V)	0.0	9.9	14.8	13.9	13.6	0.0	9.7	23.7	26.5	26.9
Gasoline mild (48 V)	0.0	2.6	4.2	4.3	3.8	0.0	4.4	9.6	10.6	10.5
Gasoline full	0.4	0.4	0.7	0.8	0.7	0.4	0.5	1.3	1.7	1.9
Gasoline PHEV	0.1	1.0	2.1	2.9	2.7	0.3	0.9	2.2	2.9	3.2
Gasoline EREV	0.0	0.1	0.2	0.3	0.3	0.1	0.2	0.4	0.5	0.6
Diesel	2.2	2.1	1.4	1.2	1.2	12.1	12.8	9.7	7.8	6.3
Diesel mild (48 V)	0.0	0.2	0.2	0.2	0.2	0.0	1.5	3.7	3.7	2.9
Diesel PHEV	0.0	0.1	0.2	0.2	0.2	0.1	0.2	0.6	0.7	0.5
CNG	0.5	0.4	0.4	0.4	0.2	0.7	0.5	0.2	0.1	0.1
LPG	0.5	0.5	0.5	0.4	0.8	0.6	0.4	0.4	0.4	0.4
EV low range	0.0	0.3	2.2	6.9	6.8	0.0	1.1	2.8	3.4	3.4
EV high range	0.0	0.0	0.3	1.0	1.0	0.0	0.2	0.9	1.2	1.2
fuel cell	0.0	0.0	0.0	0.0	0.0	0.0	0.0	0.0	0.0	0.0

Appendix 3: Distribution of Sales of New Private Vehicles, by Power Train Mode, in the Green Constraint Scenario

PC	*World*					Europe				
%	*2017*	*2020*	*2025*	*2030*	*2035*	2017	2020	2025	2030	2035
Gasoline	*79.9*	*68.5*	*47.4*	*37.8*	*33.8*	50.7	51.3	37.9	3.2	0.0
Gasoline mild (12 V)	*0.5*	*8.9*	*20.3*	*22.9*	*20.9*	0.0	6.3	18.4	33.0	30.2
Gasoline mild (48 V)	*0.2*	*5.0*	*11.7*	*13.1*	*11.9*	0.0	3.1	10.0	18.9	18.1
Gasoline full	*2.4*	*1.7*	*1.9*	*2.0*	*2.0*	2.6	1.4	1.1	1.3	1.4
Gasoline PHEV	*0.5*	*1.7*	*3.9*	*5.8*	*6.7*	0.6	2.6	4.5	8.2	9.5
Gasoline EREV	*0.2*	*0.4*	*0.7*	*1.0*	*1.2*	*0.1*	0.4	0.6	0.8	0.9
Diesel	*13.3*	*8.5*	*3.9*	*2.4*	*2.1*	*44.2*	28.3	9.6	0.6	0.0
Diesel mild (48 V)	*0.0*	*0.5*	*1.3*	*1.6*	*1.2*	*0.0*	1.6	3.2	4.2	3.4
Diesel PHEV	*0.1*	*0.1*	*0.3*	*0.4*	*0.4*	*0.1*	0.2	0.3	0.4	0.3
CNG	*1.4*	*1.7*	*2.5*	*3.6*	*4.1*	*0.6*	1.0	4.4	9.2	11.4
LPG	*0.9*	*1.1*	*1.3*	*1.5*	*1.7*	*0.7*	1.0	1.2	1.3	1.2
EV low range	*0.7*	*1.8*	*4.3*	*6.6*	*8.0*	*0.5*	2.6	7.9	16.3	19.4
EV high range	*0.0*	*0.1*	*0.6*	*1.3*	*1.9*	*0.0*	0.2	0.9	2.8	4.0
Fuel cell	*0.0*	*0.0*	*0.0*	*0.0*	*0.1*	*0.0*	0.0	0.0	0.0	0.2
PC	China					USA				
%	2017	2020	2025	2030	2035	2017	2020	2025	2030	2035
Gasoline	94.8	77.1	46.3	35.6	32.5	92.7	70.3	39.8	27.3	23.1
Gasoline mild (12 V)	1.9	11.6	23.8	24.8	22.6	0.0	10.3	24.4	24.3	21.0
Gasoline mild (48 V)	0.9	6.0	14.1	16.3	15.9	0.0	8.3	20.5	21.7	19.6
Gasoline full	0.0	0.4	0.9	1.1	1.2	3.2	2.7	3.8	4.8	5.0
Gasoline PHEV	0.3	1.4	5.0	7.0	8.5	1.0	3.0	4.8	11.1	16.0
Gasoline EREV	0.2	0.6	1.6	2.2	2.4	0.5	0.9	0.8	1.7	2.6
Diesel	0.4	0.5	0.6	0.5	0.5	0.2	0.5	1.2	1.4	1.3
Diesel mild (48 V)	0.0	0.1	0.3	0.4	0.4	0.0	0.1	0.8	1.4	1.3
Diesel PHEV	0.0	0.0	0.0	0.0	0.0	0.2	0.3	0.7	1.2	1.6
CNG	0.0	0.0	0.0	0.0	0.0	0.7	0.5	0.4	0.4	0.4
LPG	0.0	0.0	0.0	0.0	0.0	0.5	0.3	0.2	0.2	0.1

(continued)

(continued)

PC	China					USA				
%	2017	2020	2025	2030	2035	2017	2020	2025	2030	2035
EV low range	1.5	2.3	6.6	10.3	13.5	1.0	2.7	2.1	3.4	5.6
EV high range	0.0	0.1	0.8	1.8	2.5	0.0	0.1	0.5	1.2	2.3
Fuel cell	0.0	0.0	0.0	0.0	0.0	0.0	0.0	0.0	0.0	0.1
PC	Japan					India				
%	2017	2020	2025	2030	2035	2017	2020	2025	2030	2035
Gasoline	62.1	57.8	35.0	25.3	21.3	50.3	59.2	55.7	56.3	56.1
Gasoline mild (12 V)	0.0	6.5	16.3	21.0	17.2	0.0	6.3	18.8	19.4	19.5
Gasoline mild (48 V)	0.0	4.2	13.7	14.4	12.2	0.0	0.6	2.1	2.5	2.4
Gasoline full	21.3	12.6	6.9	4.6	4.3	0.0	0.1	0.1	0.1	0.1
Gasoline PHEV	0.5	1.7	4.1	4.8	5.6	0.0	0.7	2.6	2.4	1.8
Gasoline EREV	0.1	0.4	0.9	0.9	0.9	0.0	0.0	0.0	0.0	0.0
Diesel	14.6	10.8	5.1	3.7	3.1	42.9	22.0	6.5	3.5	2.5
Diesel mild (48 V)	0.0	0.9	2.9	3.3	2.8	0.0	0.6	1.2	1.2	0.8
Diesel PHEV	0.1	0.2	0.4	0.5	0.5	0.0	0.3	1.0	0.8	0.6
CNG	0.6	1.4	1.9	2.7	2.0	6.2	8.9	8.7	9.5	10.2
LPG	0.6	0.9	1.2	1.3	1.4	0.5	0.8	1.1	1.6	2.1
EV low range	0.2	2.2	9.8	13.4	19.8	0.0	0.4	1.8	2.1	3.0
EV high range	0.0	0.2	1.7	4.2	8.0	0.0	0.0	0.3	0.6	0.8
Fuel cell	0.0	0.0	0.0	0.1	0.8	0.0	0.0	0.0	0.0	0.1
PC	Brazil					Russia				
%	2017	2020	2025	2030	2035	2017	2020	2025	2030	2035
Gasoline	98.4	84.8	74.7	69.3	70.5	86.9	71.0	47.4	42.7	43.7
Gasoline mild (12 V)	0.0	10.0	15.0	14.2	13.9	0.0	9.5	24.7	27.6	27.5
Gasoline mild (48 V)	0.0	2.6	4.1	4.2	3.7	0.0	3.9	9.5	10.7	10.5
Gasoline full	0.4	0.4	0.7	0.8	0.7	0.4	0.5	1.1	1.5	1.7
Gasoline PHEV	0.1	1.0	2.0	2.8	2.6	0.2	0.9	2.0	2.6	3.0
Gasoline EREV	0.0	0.1	0.2	0.3	0.3	0.1	0.2	0.4	0.6	0.6
Diesel	0.0	0.0	0.0	0.0	0.0	10.9	11.4	9.1	7.5	6.2
Diesel mild (48 V)	0.0	0.0	0.0	0.0	0.0	0.0	1.1	3.1	3.3	2.7
Diesel PHEV	0.0	0.0	0.1	0.1	0.0	0.1	0.2	0.3	0.4	0.3

(continued)

(continued)

PC	Brazil					Russia				
%	2017	2020	2025	2030	2035	2017	2020	2025	2030	2035
CNG	0.5	0.4	0.4	0.4	0.2	0.7	0.5	0.2	0.1	0.1
LPG	0.5	0.5	0.5	0.4	0.8	0.6	0.4	0.4	0.4	0.4
EV low range	0.0	0.2	2.1	6.6	6.4	0.0	0.4	1.3	2.1	2.6
EV high range	0.0	0.0	0.3	0.9	0.8	0.0	0.1	0.3	0.6	0.8
Fuel cell	0.0	0.0	0.0	0.0	0.0	0.0	0.0	0.0	0.0	0.0

Appendix 4: Distribution of Sales of New Light Commercial Vehicles, by Power Train Mode, in the Green Constraint Scenario

LCV	*World*					Europe				
%	*2017*	*2020*	*2025*	*2030*	*2035*	2017	2020	2025	2030	2035
Gasoline	*49.3*	*42.0*	*24.7*	*18.6*	*18.3*	8.5	14.6	18.6	2.0	0.0
Gasoline mild (12 V)	*0.6*	*7.3*	*13.2*	*14.2*	*13.3*	0.0	3.3	12.1	20	18.0
Gasoline mild (48 V)	*0.2*	*3.8*	*7.3*	*8.3*	*8.2*	0.0	1.7	6.4	10.9	10.3
Gasoline full	*0.1*	*0.4*	*1.1*	*1.5*	*1.7*	0.0	0.1	0.2	0.3	0.3
Gasoline PHEV	*0.3*	*2.4*	*4.9*	*5.9*	*6.8*	0.5	2.4	4.0	6.7	7.5
Gasoline EREV	*0.0*	*0.0*	*0.0*	*0.0*	*0.0*	0.0	0.0	0.0	0.0	0.0
Diesel	*47.0*	*31.9*	*12.3*	*6.8*	*6.2*	90.1	67.3	26.0	1.6	0.0
Diesel mild (48 V)	*0.5*	*3.7*	*5.8*	*5.9*	*4.9*	0.0	4.9	10.4	14	11.6
Diesel PHEV	*0.9*	*2.7*	*5.2*	*6.8*	*6.2*	0.4	2.1	4.1	6.1	6.1
CNG	*0.2*	*0.3*	*1.1*	*2.5*	*3.1*	0.0	0.3	3.2	10.6	14.5
LPG	*0.7*	*0.8*	*1.1*	*1.4*	*1.7*	0.1	0.2	0.4	0.6	0.6
EV low range	*0.1*	*2.8*	*13.2*	*17.3*	*18.7*	0.3	2.5	10.7	18.6	20.3

(continued)

(continued)

LCV	*World*					Europe				
%	*2017*	*2020*	*2025*	*2030*	*2035*	2017	2020	2025	2030	2035
EV high range	*0.0*	*2.0*	*10.1*	*10.9*	*10.9*	0.0	0.4	3.9	8.6	10.8
Fuel cell	*0.0*	*0.0*	*0.0*	*0.0*	*0.0*	0.0	0.0	0.0	0.0	0.0
LCV	China					USA				
%	2017	2020	2025	2030	2035	2017	2020	2025	2030	2035
Gasoline	56.4	46.7	21.7	13.7	11.4	36.8	33.2	23.0	15.5	11.7
Gasoline mild (12 V)	1.4	7.1	11.9	10.1	8.7	0.0	11.3	18.8	16.2	12.5
Gasoline mild (48 V)	0.5	2.8	5.7	6.3	6.4	0.0	10.9	19.3	17.8	14.2
Gasoline full	0.0	0.2	0.4	0.4	0.4	0.8	2.0	5.5	7.7	7.1
Gasoline PHEV	0.0	2.6	5.4	4.9	5.4	0.9	4.1	6.2	12.0	15.9
Gasoline EREV	0.0	0.0	0.0	0.0	0.0	0.0	0.0	0.0	0.0	0.0
Diesel	39.5	28.2	7.3	3.6	3.2	59.7	30.6	11.1	6.4	5.1
Diesel mild (48 V)	1.1	3.9	5.0	3.9	3.4	0.0	3.7	7.1	7.0	6.0
Diesel PHEV	1.0	3.5	7.8	11.9	11.7	0.8	2.1	3.0	4.8	6.4
CNG	0.0	0.0	0.0	0.0	0.0	0.4	0.5	0.5	0.5	0.6
LPG	0.0	0.0	0.0	0.0	0.0	0.5	0.4	0.3	0.2	0.2
EV low range	0.0	1.5	15.4	22.9	26.3	0.0	1.2	4.4	9.3	15.1
EV high range	0.0	3.5	19.4	21.8	23.0	0.0	0.1	0.8	2.6	5.2
Fuel cell	0.0	0.0	0.0	0.0	0.0	0.0	0.0	0.0	0.0	0.1
LCV	Japan					India				
%	2017	2020	2025	2030	2035	2017	2020	2025	2030	2035
Gasoline	88.1	56.2	20.1	12.3	10.1	54.4	44.8	24.8	23.5	21.6
Gasoline mild (12 V)	0.0	7.4	12.9	11.7	9.7	0.0	6.6	14.9	15.7	14.5
Gasoline mild (48 V)	0.0	6.2	10.5	9.2	8.2	0.0	1.4	3.1	3.5	3.3
Gasoline full	0.7	1.6	2.4	3.2	3.3	0.0	0.0	0.0	0.0	0.0
Gasoline PHEV	0.8	2.2	4.1	6.3	7.4	0.0	2.4	7.4	6.9	7.3

(continued)

(continued)

LCV	Japan					India				
%	2017	2020	2025	2030	2035	2017	2020	2025	2030	2035
Gasoline EREV	0.0	0.0	0.0	0.0	0.0	0.0	0.0	0.0	0.0	0.0
Diesel	8.5	7.3	7.7	8.9	7.0	44.1	29.9	11.6	9.0	6.9
Diesel mild (48 V)	0.0	1.6	4.9	7.3	6.3	0.0	1.5	3.0	3.2	2.7
Diesel PHEV	0.9	1.6	2.6	4.0	4.2	0.0	1.8	5.1	4.8	4.9
CNG	0.0	0.0	0.0	0.0	0.0	0.7	0.9	0.6	0.5	0.4
LPG	0.6	0.6	0.3	0.2	0.3	0.5	0.5	0.4	0.6	0.7
EV low range	0.4	13.2	26.6	25.6	27.4	0.2	9.6	25.4	26.5	30.1
EV high range	0.0	2.2	7.9	11.1	16.2	0.0	0.7	3.7	5.9	7.6
Fuel cell	0.0	0.0	0.0	0.0	0.0	0.0	0.0	0.0	0.0	0.1
LCV	Brazil					Russia				
%	2017	2020	2025	2030	2035	2017	2020	2025	2030	2035
Gasoline	31.2	30.7	32.8	30.2	28.9	66.5	40.6	19.7	15.9	16.3
Gasoline mild (12 V)	0.0	8.1	8.9	8.2	7.6	0.0	10.9	15.6	15.3	15.5
Gasoline mild (48 V)	0.0	4.4	5.1	5.1	4.9	0.0	7.9	10.6	10.5	10.8
Gasoline full	0.1	0.0	0.1	0.1	0.1	0.6	0.7	2.7	4.2	4.8
Gasoline PHEV	0.4	1.0	3.2	5.0	5.5	0.7	0.9	3.6	6.2	7.2
Gasoline EREV	0.0	0.0	0.0	0.0	0.0	0.0	0.0	0.0	0.0	0.0
Diesel	67.2	49.0	34.0	25.9	24.4	29.6	23.6	14.5	10.7	8.5
Diesel mild (48 V)	0.0	3.6	4.5	4.4	4.3	0.0	5.1	8.7	8.0	6.4
Diesel PHEV	0.4	0.9	2.4	3.3	3.5	0.7	0.9	3.0	4.2	4.0
CNG	0.4	0.6	0.8	0.7	0.6	1.3	0.5	0.1	0.0	0.0
LPG	0.3	0.5	0.7	0.7	0.8	0.6	0.3	0.2	0.2	0.2
EV low range	0.0	1.0	5.9	13.0	14.7	0.0	7.2	15.7	17.1	17.6
EV high range	0.0	0.2	1.7	3.4	4.7	0.0	1.4	5.6	7.7	8.6
Fuel cell	0.0	0.0	0.0	0.0	0.1	0.0	0.0	0.0	0.0	0.0

Appendix 5: Distribution of Sales of New Light Vehicles by Power Train Mode in 2030 and 2035, in the 4 Scenarios

LV	World 2030				World 2035			
%	Liberal world	Stagnation	Green growth	Green con-straint	liberal world	Stagnation	green growth	Green con-straint
Gasoline	34.3	39.2	28.5	35.7	33.5	40.1	28.6	35.8
Gasoline mild (12 V)	22.2	25.2	16.4	21.9	20.1	24.2	14.9	20.1
Gasoline mild (48 V)	13.9	14.3	10.1	12.6	12.9	13.8	8.9	11.5
Gasoline Full	2.2	2.0	1.9	1.9	2.3	2.2	1.7	2.0
Gasoline PHEV	5.8	3.5	8.6	5.8	7.3	3.9	9.1	6.7
Gasoline EREV	0.8	0.4	1.8	0.9	1.0	0.4	1.9	1.1
Diesel	3.5	3.4	2.7	2.9	3.6	3.0	2.6	2.5
Diesel Mild (48 V)	2.6	2.6	1.7	2.0	2.4	2.2	1.4	1.6
Diesel PHEV	1.3	0.9	1.3	1.1	1.4	0.8	1.2	1.0
CNG	4.0	2.6	4.6	3.4	4.2	2.8	4.6	4.0
LPG	1.9	1.3	1.9	1.5	2.1	1.4	1.9	1.7
EV low range	5.7	3.3	13.5	7.8	7.0	4.0	15.5	9.1
EV high range	1.7	0.9	5.4	2.4	2.2	1.0	7.5	2.8
Fuel cell	0.0	0.0	0.2	0.0	0.1	0.0	0.2	0.1

Appendix 6: Distribution of Sales of New Light Vehicles by Power Train Mode in France

%	France–LCV					France–PC					France–LV				
	2017	2020	2025	2030	2035	2017	2020	2025	2030	2035	2017	2020	2025	2030	2035
Gasoline	5.3	13.3	18.1	1.9	0.0	44.3	37.4	23.3	1.4	0.0	38.8	34.0	22.6	1.5	0.0
Gasoline mild (12 V)	0.0	3.3	13.5	21.6	18.9	0.0	4.8	11.8	19.8	17.1	0.0	4.6	12.0	20.1	17.3
Gasoline mild (48 V)	0.0	1.5	6.8	11.5	11.0	0.0	2.3	7.5	13.8	12.6	0.0	2.2	7.4	13.5	12.3
Gasoline full	0.0	0.0	0.1	0.1	0.2	2.7	1.4	0.8	0.6	0.5	2.3	1.2	0.7	0.5	0.5
Gasoline PHEV	0.4	1.6	3.4	5.9	7.3	0.8	5.0	11.8	18.5	20.7	0.8	4.5	10.6	16.8	18.8
Gasoline EREV	0.0	0.0	0.0	0.0	0.0	0.2	0.7	1.2	1.1	1.0	0.1	0.6	1.0	1.0	0.9
Diesel	92.6	71.1	26.0	1.3	0.0	49.9	38.7	17.3	0.7	0.0	55.9	43.2	18.5	0.7	0.0
Diesel mild (48 V)	0.0	4.5	11.2	13.2	9.8	0.0	2.3	4.9	5.6	3.9	0.0	2.6	5.8	6.6	4.8
Diesel PHEV	0.4	1.5	4.0	5.6	5.8	0.1	0.2	0.4	0.4	0.3	0.1	0.4	0.9	1.1	1.1
CNG	0.0	0.2	2.7	9.8	13.1	0.2	0.6	3.5	8.3	9.7	0.2	0.5	3.4	8.5	10.2
LPG	0.0	0.0	0.0	0.0	0.0	0.0	0.0	0.0	0.0	0.0	0.0	0.0	0.0	0.0	0.0
EV low range	1.3	2.5	10.5	20.1	22.0	1.8	6.4	16.0	26.3	29.2	1.8	5.8	15.2	25.4	28.2
EV high range	0.0	0.5	3.7	9.0	11.9	0.0	0.3	1.6	3.4	4.6	0.0	0.4	1.9	4.2	5.6

Appendix 7: Distribution of Sales of New HV, by Power Train Mode, in the Green Constraint Scenario

HV	*World*					Europe				
%	2017	2020	2025	*2030*	*2035*	2017	2020	2025	2030	2035
Diesel	*99.9*	*99.0*	*95.3*	*87.2*	*83.1*	*99.3*	91.5	67.2	5.9	0.0
Diesel full hybrid series	*0.0*	*0.0*	*0.3*	*0.7*	*0.8*	*0.0*	0.1	0.6	1.7	2.3
Diesel full hybrid parallel	*0.0*	*0.2*	*0.8*	*4.9*	*4.6*	*0.0*	1.4	5.7	49.1	48.9
Diesel plug-in hybrid parallel	*0.0*	*0.0*	*0.1*	*0.3*	*0.4*	*0.0*	0.1	0.8	2.2	3.2
CNG	*0.0*	*0.0*	*0.0*	*0.1*	*0.1*	0.0	0.0	0.3	0.7	1.5
LNG	*0.0*	*0.0*	*0.0*	*0.1*	*0.2*	0.0	0.0	0.3	0.8	1.9
BEV	*0.1*	*0.8*	*3.4*	*6.8*	*10.7*	0.6	6.9	25.2	39.6	42.1
Fuel cell	*0.0*	*0.0*	*0.0*	*0.0*	*0.0*	0.0	0.0	0.0	0.0	0.0
HV	China					USA				
%	2017	2020	2025	2030	2035	2017	2020	2025	2030	2035
Diesel	99.3	91.5	67.2	5.9	0.0	99.9	99.7	96.4	84.6	74.2
Diesel full hybrid series	0.0	0.1	0.6	1.7	2.3	0.0	0.0	0.0	0.1	0.1
Diesel full hybrid parallel	0.0	1.4	5.7	49.1	48.9	0.0	0.0	0.4	0.8	1.1
Diesel plug-in hybrid parallel	0.0	0.1	0.8	2.2	3.2	0.0	0.0	0.0	0.2	0.3
CNG	0.0	0.0	0.3	0.7	1.5	0.0	0.0	0.0	0.1	0.1
LNG	0.0	0.0	0.3	0.8	1.9	0.0	0.0	0.1	0.1	0.1
BEV	0.6	6.9	25.2	39.6	42.1	0.0	0.2	3.1	14.2	24.0
Fuel cell	0.0	0.0	0.0	0.0	0.0	0.0	0.0	0.0	0.0	0.0
HV	Japan					India				
%	2017	2020	2025	2030	2035	2017	2020	2025	2030	2035
Diesel	100.0	99.8	96.8	91.2	87.0	100.0	100.0	99.6	97.3	91.5
Diesel full hybrid series	0.0	0.0	1.2	3.1	3.8	0.0	0.0	0.1	0.9	2.0
Diesel full hybrid parallel	0.0	0.1	1.4	3.2	3.8	0.0	0.0	0.0	0.2	0.3
Diesel plug-in hybrid parallel	0.0	0.0	0.1	0.2	0.4	0.0	0.0	0.0	0.1	0.2
CNG	0.0	0.0	0.0	0.0	0.0	0.0	0.0	0.0	0.0	0.0
LNG	0.0	0.0	0.0	0.0	0.0	0.0	0.0	0.0	0.0	0.0
BEV	0.0	0.0	0.5	2.3	5.1	0.0	0.0	0.2	1.6	6.0
Fuel cell	0.0	0.0	0.0	0.0	0.0	0.0	0.0	0.0	0.0	0.0

(continued)

(continued)

HV	Brazil					Russia				
%	2017	2020	2025	2030	2035	2017	2020	2025	2030	2035
Diesel	100.0	100.0	99.9	98.8	96.3	100.0	99.8	99.1	96.0	91.0
Diesel full hybrid series	0.0	0.0	0.0	0.1	0.3	0.0	0.0	0.0	0.2	0.3
Diesel full hybrid parallel	0.0	0.0	0.1	0.4	1.1	0.0	0.0	0.0	0.2	0.4
Diesel plug-in hybrid parallel	0.0	0.0	0.0	0.0	0.0	0.0	0.0	0.0	0.0	0.0
CNG	0.0	0.0	0.0	0.0	0.0	0.0	0.0	0.1	0.1	0.1
LNG	0.0	0.0	0.0	0.0	0.0	0.0	0.1	0.4	0.6	0.7
BEV	0.0	0.0	0.1	0.6	2.4	0.0	0.0	0.4	2.8	7.5
Fuel cell	0.0	0.0	0.0	0.0	0.0	0.0	0.0	0.0	0.0	0.0

Appendix 8: Distribution of Sales of New HV by Power Train Mode in 2030 and 2035, in the 4 Scenarios

HV	World 2030				World 2035			
%	Liberal world	Stagnation	Green growth	Green constraint	Liberal world	Stagnation	Green growth	Green constraint
Diesel	87.9	90.1	79.2	87.2	82.5	88.6	66.2	83.1
Diesel full hybrid series	0.7	0.5	0.9	0.7	0.9	0.7	1.1	0.8
Diesel full hybrid parallel	4.4	5.4	3.8	4.9	3.3	5.0	2.4	4.6
Diesel plug-in hybrid parallel	0.3	0.2	0.3	0.3	0.4	0.3	0.5	0.4
CNG	0.5	0.0	1.1	0.1	1.1	0.0	2.1	0.1
LNG	0.2	0.0	0.4	0.1	0.5	0.1	0.7	0.2
BEV	5.9	3.7	14.4	6.8	11.1	5.3	26.6	10.7
Fuel cell	0.0	0.0	0.0	0.0	0.2	0.0	0.4	0.0

Appendix 9: Distribution of Sales of New Fuel Cell Light Vehicles According to Hydrogen Prices at the Pump

Using the Green Constraint scenario different prices per kg of hydrogen at the pump, reached in the model in 2030, were tested:

- Green Constraint—Reference scenario
- "GC €7"—Green Constraint–7 Euros/kg(H_2)
- "GC €6"—Green Constraint–6 Euros/kg(H_2)
- "GC €5"—Green Constraint–5 Euros/kg(H_2)
- "GC €4"—Green Constraint–4 Euros/kg(H_2)

LV	*World*		Europe		China		NAFTA		Japan + South Korea	
%	*2030*	*2035*	2030	2035	2030	2035	2030	2035	2030	2035
Green constraint	*<0.1*	*0.1*	<0.1	0.2	<0.1	<0.1	<0.1	0.1	<0.1	0.4
GC €7	*<0.1*	*0.3*	0.1	1.0	<0.1	0.1	<0.1	0.1	0.1	1.0
GC €6	*0.1*	*0.5*	0.2	2.0	<0.1	0.1	<0.1	0.2	0.1	1.7
GC €5	*0.2*	*1.1*	0.6	4.7	<0.1	0.3	<0.1	0.7	0.3	3.6
GC €4	*0.2*	*1.8*	1.0	6.9	0.1	0.7	0.1	1.2	0.5	5.2

Appendix 10: Optimum European Powertrain Mix to Achieve Different CAFE Targets by 2030 from the Green Constraint Scenario

Based on the forecasts of the Green Constraint scenario, a methodology has been developed to evaluate the mixes that allow for the optimal achievement of more restrictive average emission targets by 2030. This optimality is defined by calculating step by step the market strategy that allows the greatest decrease in average CO_2 emissions with the least difference in cost compared to the base case of the baseline scenario (here the Green Constraint scenario).

Powertrain	Market shares (PC)					
	2021 (%)	2030—base case (%)	2030—target–30% (%)	2030—Target–35% (%)	2030—Target–40% (%)	2030—Target–45% (%)
Gasoline	50.1	3.2	3.0	2.7	2.5	2.2
Gasoline mild (12 V)	8.9	33.0	29.4	26.4	23.4	20.5
Gasoline mild (48 V)	4.4	18.9	17.5	16.1	14.5	12.6
Gasoline full	1.3	1.3	1.0	1.0	0.9	0.8
Gasoline PHEV	3.0	8.2	9.1	10.1	12.3	15.2
Gasoline EREV	0.4	0.8	0.7	0.7	0.8	0.8
Diesel	23.2	0.6	0.5	0.5	0.4	0.4
Diesel Mild (48 V)	2.1	4.2	4.0	3.8	3.4	3.0
Diesel PHEV	0.2	0.4	0.4	0.5	0.6	0.7
CNG	1.4	9.2	8.2	7.9	7.4	7.0
LPG	1.1	1.3	1.1	1.0	1.0	0.9
BEV	3.7	16.3	19.8	22.5	25.3	27.8
BEV300	0.3	2.8	5.2	6.7	7.3	8.1
Fuel cell	0.0	0.0	0.1	0.1	0.1	0.1
Average emissions (WLTP—g/km)	121.9	79.2	73.0	68.2	63.1	57.6

(continued)

(continued)

Powertrain	Market shares (LCV)					
	2021 (%)	2030—base case (%)	2030—Target–30% (%)	2030—Target–35% (%)	2030—Target–40% (%)	2030—Target–45% (%)
Gasoline	16.9	2.0	2.02	2.02	2.02	2.02
Gasoline mild (12 V)	5.0	20.0	19.96	19.96	19.96	19.96
Gasoline mild (48 V)	2.5	10.9	10.88	10.88	10.88	10.88
Gasoline full	0.1	0.3	0.30	0.30	0.30	0.30
Gasoline PHEV	2.8	6.7	6.67	6.67	6.67	6.67
Gasoline EREV	0.0	0.0	0.00	0.00	0.00	0.00
Diesel	58.2	1.6	1.57	1.57	1.57	1.57
Diesel Mild (48 V)	6.3	14.0	14.00	14.00	14.00	14.00
Diesel PHEV	2.6	6.1	6.15	6.15	6.15	6.15
CNG	0.6	10.6	10.64	10.64	10.64	10.64
LPG	0.2	0.6	0.58	0.58	0.58	0.58
BEV	3.9	18.6	18.62	18.62	18.62	18.62
BEV300	0.9	8.6	8.63	8.63	8.63	8.63
Fuel cell	0.0	0.0	0.00	0.00	0.00	0.00
Average emissions (WLTP—g/km)	153.5	85.6	85.6	85.6	85.6	85.6

Appendix 11: Demand for Fuel and Electricity for Each Area of LV and HV, Green Constraint Scenario

LV and HV	*World*					Europe				
Mt (excluding electricity in TWh)	*2017*	*2020*	*2025*	*2030*	*2035*	2017	2020	2025	2030	2035
Gasoline (Mt)	*931.4*	*945.0*	*927.9*	*878.7*	*825.2*	49.9	44.4	41.8	38.4	32.5
Diesel (Mt)	*968.3*	*1006.6*	*1020.9*	*1010.4*	*994.4*	226.8	227.5	206.5	169.6	132.4
Bioethanol (Mt)	*95.7*	*101.0*	*110.3*	*121.7*	*129.0*	3.8	4.0	5.1	6.9	7.9
Biodiesel (Mt)	*32.2*	*37.3*	*44.8*	*51.7*	*58.4*	13.9	14.7	14.6	13.0	10.9
CNG (Mt)	*22.8*	*21.9*	*20.7*	*21.5*	*24.6*	0.5	0.5	1.1	2.8	4.9
LNG (Mt)	*0.3*	*0.3*	*0.4*	*0.7*	*1.9*	<0.1	<0.1	<0.1	0.5	1.8
LPG (Mt)	*23.2*	*25.4*	*30.3*	*35.5*	*39.6*	1.7	1.4	1.2	1.0	1.0
Hydrogen (Mt)	*<0.1*	*<0.1*	*<0.1*	*<0.1*	*0.2*	<0.1	<0.1	<0.1	<0.1	<0.1
Electricity (TWh)	*43.2*	*89.2*	*275.9*	*555.8*	*881.5*	0.9	6.1	41.9	112.7	186.7
LV and HV	China					USA				
Mt (excluding electricity in TWh)	2017	2020	2025	2030	2035	2017	2020	2025	2030	2035
Gasoline (Mt)	150.9	176.2	192.9	182.0	160.0	288.4	257.2	217.7	181.3	151.6
Diesel (Mt)	142.7	156.8	158.7	146.8	131.6	144.2	142.6	141.3	138.5	132.9
Bioethanol (Mt)	6.4	10.6	17.7	24.5	28.8	51.8	62.6	75.4	82.5	81.2
Biodiesel (Mt)	0.6	1.0	1.8	2.3	2.7	3.7	4.9	6.8	8.6	9.7
CNG (Mt)	1.3	1.1	0.7	0.3	<0.1	1.6	1.6	1.4	1.0	0.7
LNG (Mt)	<0.1	<0.1	<0.1	<0.1	<0.1	0.3	0.3	0.2	0.1	0.1
LPG (Mt)	<0.1	<0.1	<0.1	<0.1	<0.1	2.1	2.1	1.8	1.3	0.8
Hydrogen (Mt)	<0.1	<0.1	<0.1	<0.1	<0.1	<0.1	<0.1	<0.1	<0.1	<0.1
Electricity (TWh)	25.6	36.5	92.4	175.2	253.4	1.6	6.5	17.0	35.7	71.8
LV and HV	Japan					India				
Mt (excluding electricity in TWh)	2017	2020	2025	2030	2035	2017	2020	2025	2030	2035
Gasoline (Mt)	26.6	23.7	19.1	15.1	11.6	21.3	26.0	29.5	31.7	36.0
Diesel (Mt)	22.1	21.9	20.8	19.3	17.7	65.2	74.6	87.4	100.7	108.4
Bioethanol (Mt)	0.3	0.5	0.7	0.8	0.8	0.9	1.7	3.0	4.5	6.1
Biodiesel (Mt)	<0.1	<0.1	<0.1	<0.1	<0.1	2.0	4.6	9.7	16.2	21.7
CNG (Mt)	<0.1	0.1	0.1	0.2	0.2	1.3	1.3	1.4	1.6	2.1
LNG (Mt)	<0.1	<0.1	<0.1	<0.1	<0.1	<0.1	<0.1	<0.1	<0.1	<0.1
LPG (Mt)	0.9	0.6	0.5	0.5	0.5	1.2	1.0	0.9	0.9	1.4
Hydrogen (Mt)	<0.1	<0.1	<0.1	<0.1	<0.1	<0.1	<0.1	<0.1	<0.1	<0.1
Electricity (TWh)	0.3	0.9	5.5	11.1	17.6	2.1	7.0	28.5	58.4	93.7

(continued)

(continued)

LV and HV	Brazil					Russia				
Mt (excluding electricity in TWh)	2017	2020	2025	2030	2035	2017	2020	2025	2030	2035
Gasoline (Mt)	21.3	26.0	29.5	31.7	36.0	41.5	41.4	41.7	42.4	42.3
Diesel (Mt)	65.2	74.6	87.4	100.7	108.4	14.1	14.7	15.8	16.5	16.3
Bioethanol (Mt)	0.9	1.7	3.0	4.5	6.1	<0.1	<0.1	<0.1	<0.1	<0.1
Biodiesel (Mt)	2.0	4.6	9.7	16.2	21.7	<0.1	<0.1	<0.1	<0.1	<0.1
CNG (Mt)	1.3	1.3	1.4	1.6	2.1	<0.1	0.1	0.1	0.1	0.1
LNG (Mt)	<0.1	<0.1	<0.1	<0.1	<0.1	<0.1	<0.1	<0.1	<0.1	<0.1
LPG (Mt)	1.2	1.0	0.9	0.9	1.4	6.3	5.8	5.0	4.2	2.6
Hydrogen (Mt)	<0.1	<0.1	<0.1	<0.1	<0.1	<0.1	<0.1	<0.1	<0.1	<0.1
Electricity (TWh)	2.1	7.0	28.5	58.4	93.7	<0.1	0.2	1.8	4.6	8.1

Automotive Meets ICT—Enabling the Shift of Value Creation Supported by European R&D

Eric Armengaud, Bernhard Peischl, Peter Priller and Omar Veledar

Abstract Digitalization is proving to be a game changer in bridging the gap between the heterogeneous skills and markets. It increases productivity through optimisation over the entire supply chain and lets new services emerge through the convergence of applications domains. In this paper, we are providing a review of the main automotive trends and are highlighting how digitalization (especially by information communication technologies—ICT) is supporting, even pushing innovation. We are especially mapping to the IOT4CPS and SCOTT projects to present key results related to Internet of Things supporting the digital transition in the automotive domain.

Keywords Automotive · Digitalization · CPS · IOT · SCOTT · IOT4CPS

1 Introduction

Digitalization is a game changer in bridging the gap between the heterogeneous skills and markets. Firstly, digitalization supports the increase of productivity through optimization over the entire supply chain by seamlessly connecting the various stages of the product lifecycle. This enables an earlier and more accurate prediction of the real behavior during a concept phase (frontloading) as well as faster and more tailored improvement of development and production processes taking into account real product's usage. Secondly, digitalization enables the emergence of new services through the convergence of different applications domains (Fig. 1).

E. Armengaud (✉) · B. Peischl · P. Priller · O. Veledar
Research & Technology Development, AVL List GmbH, Graz, Austria
e-mail: eric.armengaud@avl.com

B. Peischl
e-mail: bernhard.peischl@avl.com

P. Priller
e-mail: peter.priller@avl.com

O. Veledar
e-mail: omar.veledar@avl.com

J. Langheim (ed.), *Electronics Components and Systems for Automotive Applications*, Lecture Notes in Mobility,
https://doi.org/10.1007/978-3-030-14156-1_4

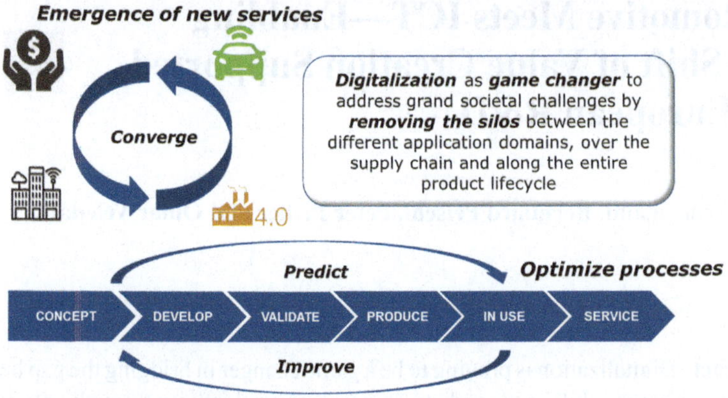

Fig. 1 Digitalization as game changer

This is especially true for the automotive domain. Nowadays, the automotive sector is confronted to four main trends:

- **Electrification** (ETPs 2017), with the introduction of e-mobility (hybrid, pure electric vehicle) to optimize or even completely remove the internal combustion engine, finally reducing the resulting pollutant emissions during vehicle operation
- **ADAS and autonomous driving functions** (ERTRAC 2017a), with the purpose of providing more comprehensive information to the driver for better context awareness, up to taking over specific driving maneuvers—finally reducing the demands on the driver and reducing number and impact of possible accidents
- **Connected vehicles** (McKinsey and Company 2016) enabling optimization of vehicle's operation or the emergence of new services while relying on external information, e.g. from other vehicles or from the infrastructure
- **Diverse mobility** (2017b) targeting the efficient movement of people and goods with respect to different factors such that time, energy consumption, ecological footprint.

For all the trends, digitalization is playing a pivotal role as supporting technology (e.g., power electronics and smart control for electrification) as well as enabling technology. This is evident for autonomous driving with the development and deployment of smart systems for advanced environment sensing and complex decision making supported by cyber-physical systems, for connectivity where Internet of Things is playing a key role for optimized machine to machine communication, or for diverse mobility taking advantage of the organization of complex and fairly independent data sets from different domains to finally enable the emergence of new services.

The contribution of this paper is a proposal to structure the technology stack, therefore making the link between value creation and required expertise. The remaining of the paper is organized as follow: Sect. 2 discusses the challenges of deploying market specific end-to-end ICT stacks; Sects. 3 and 4 are providing an overview of the R&D projects SCOTT[1] and IOT4CPS[2] as respectively large European and Austrian initiatives for Internet of Things, thus illustrating main research domains. Finally, Sect. 5 concludes this work.

2 The ICT Stack

The size of the software in software-intensive systems follows Moore's law and increases with an order of magnitude every 5–10 years, depending on the domain. In a modern premium car, software amounts to overall more than 100 million object code instructions, totaling close to 1 GB of software (Ebert and Jones 2009). With this trend being continued, the value creation in and around cars is increasingly determined by embedded software and data. This is driven by three main trends.

Firstly, due to the large-scale adoption (e.g., consider the mobile phone market), the cost of sensors and actuators has dropped dramatically over the last decade (Hatler et al. 2012). Due to this inexpensive and wide-scale deployment of sensors and actuators, today's ICT systems obtain orders of magnitude more data about the environment in which these systems operate, when compared to earlier days. Secondly, the cost of data storage has dropped radically, making it possible to store and analyze this data lake to an unprecedented extent. Finally, due to the increasing amount of computational power available to software-intensive systems, we manage to analyze data and to make considerably more dynamic, contextual decisions as compared to earlier days. Although there are promising application fields (e.g., trends such as electrification, ADAS, connected vehicles and diverse mobility in the automotive sector), there is no established methodology for systematically building homogenous end-to-end solutions.

Figure 2 proposes a high-level overview of a technology stack for data-driven automotive applications: (a) the connected devices (or **cyber-physical systems** CPS), implementing digitalization at the edge, will be in charge to interact with the physical world in order to locally digitalize relevant information and act on the environment. They are typically specialized for a dedicated application and market. (b) The **data platforms and analytics** will target the data gathering, organization processing, as well as communication management. These layers typically present a high degree of standardization to enable higher number of participants to collaborate. (c) The **market-specific application**, on top, will provide the value creation for the customer while relying on the lower layers. This layer is typically tailored for a specific market need. (d) The stack is further complemented by **data management**, including security (privacy, confidentiality) and ethics, as well as interfaces to interact with external **data markets and digital ecosystems.**

[1] https://scottproject.eu/.

[2] https://iot4cps.at/.

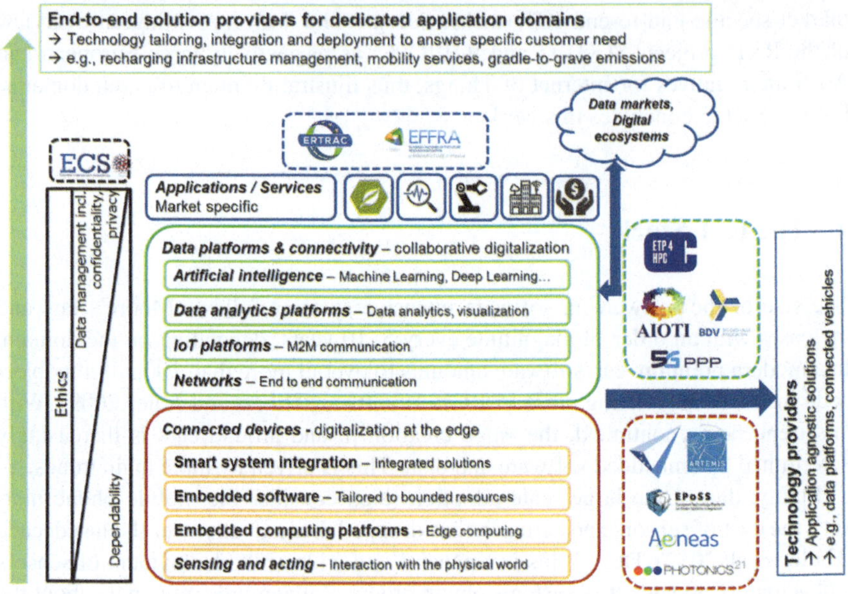

Fig. 2 Technology stack

2.1 *Cyber-Physical Systems—Digitalization at the Edge*

Cyber-physical systems (CPS) focus on the digitalization at the edge. CPS are responsible to interact with the physical world through advanced sensing, local (power-aware) processing and acting. CPS encompass different aspects:

- Sensing and acting: Refers to the capability to interact with the environment by accurate measurements of physical behaviors and actuations thereof (e.g., photonics).
- Embedded computing platforms: Denotes the capability for local processing, making a trade-off between power awareness, and performance (e.g., computing power, memory size, bandwidth).
- Embedded software: Refers to the capability to develop a software optimized for limited computing resources, low memory footprint and low energy consumption.
- Smart systems integration & dependability: Denotes the capability to deploy and tailor CPS for dedicated industrial applications, while taking into account safety, security, reliability, and timing requirements.

Important European initiatives addressing the abovementioned aspects are ECSEL[3] as partnership between Aeneas, EPoSS[4] and ARTEMIS-IA,[5] as well as the European technology platform Photonics21.[6]

2.2 *Data Platforms and the IoT—Collaborative Digitalization*

The complementary capability to edge computing is the capability to coordinate information streams and activities, therefore leveraging the local smart devices with other smart devices—cloud computing, centralized systems with higher performances, and/or by interfacing with expert knowledge. Collaborative digitalization relies on the following pillars:

- Networks: Networks need to provide the capability to exchange information in a power- aware and performance-aware manner.
- Machine to machine connectivity (Internet of Things): Refers to the capability to organize information exchange between heterogenous components and to implement functions on distributed systems.
- Data analytics, visualization: Denotes the capability to extract knowledge out of large data sets, e.g., through interaction with experts. Note that high-performance computing is playing a key role here to provide the computing resources for processing huge amounts of (often poorly structured) data.
- Data management: With the need to comply to General Data Protection Regulation[7] [GDPR, (Miche et al. 2018)], the capability to appropriately manage the data from an ethical point of view taking into account confidentiality and privacy aspects is gaining importance.

Important European initiatives are the 5G Public Private Partnership (5G-PPP[8]) for connectivity, the Alliance for Internet of Things Innovation (AIOTI) for the internet of things, the BIG Data Value Association (BDVA[9]) for big data, and the European Technology Platform for High Performance Computing (ETP4HC[10]) for high-performance computing.

---

[3]https://www.ecsel.eu/.

[4]http://www.smart-systems-integration.org/.

[5]https://artemis-ia.eu/.

[6]https://www.photonics21.org/.

[7]https://eur-lex.europa.eu/eli/reg/2016/679/oj.

[8]https://5g-ppp.eu/.

[9]http://www.bdva.eu/.

[10]http://www.etp4hpc.eu/.

2.3 Market-Specific Applications and-Services

The concepts sketched in the previous two sections build up the technology stack for innovation and differentiation on the market as these technologies enable new use cases and services in the automotive context. However, considering companies such as Tesla or Uber, none of these companies disrupted the market because they have used better or more sophisticated technology when compared to their competitors. Instead, Tesla and Uber understood the unexpressed needs from their customers and came up with innovative services to meet these needs. Thus, identifying and meeting such needs is a pivotal issue. As successful innovations are often developed before customers even express the needs that the innovation addresses (Bosch-Sijtsema and Bosch 2015), tools for collection and analysis of customer data are critical for customer-driven innovation. The ability to innovate is determined by skill sets in three pillars (Bosch 2017):

- Speed: Fast feedback cycles have the potential to outperform any other efficiency improvements. Speed in this context refers to the ability to convert an identified customer need into a new service or product feature and thus addresses the entire chain: from the identification of the customer needs, to design, implementation and maintenance.
- Data: Effective use of data from customers and products in the field is the next area to exploit and monetize. However, collecting huge amount of data is not particularly useful. Rather it is the information that can be extracted from such a data lake that ultimately has value. Extracting information from large and heterogenous volumes of data requires the use of cutting-edge software technologies, such as data mining and data analytics. Innovation comes from combining these technologies together with domain- and market knowledge to create new or deeper insights into value creation opportunities.
- Eco-systems: Strategic use of the ecosystem around the system/service is critical as this allows for agility, risk sharing and to focus on key differentiating functionality. Strategic collaboration with partners helps to offload commodity functions to ecosystem partners and in identifying differentiating services or product features.

Mastering skills related to these three pillars offers numerous opportunities to optimize development and maintenance processes (e.g., a digital-twin for a powertrain) and for the provision of new services (e.g., multimodal mobility services). However, to provide homogenous end-to-end solutions, the automotive industry needs to address the fragmentation on various levels: at the edge (CPS), at the level of IoT and data analytics as well as at the service- and application level.

Regarding automotive applications, European roadmaps are related to the European Road Transport Research Advisory Council ERTRAC[11] (ERTRAC 2017a; McKinsey and Company 2016) for road transport evolution, and the European Factories of the Future Research Association (EFFRA[12]) for smart production.

---

[11] https://www.ertrac.org/.

[12] https://www.effra.eu/.

3 The SCOTT Project

Project SCOTT boldly aims to "build trust in the Internet of Things". Started 2017 as a H2020/ECSEL Innovation Action (IA) by an European consortium of more than 50 industry and research partners, SCOTT's acronym is a good summary of the main objectives: "Secure Connected Trustable Things". SCOTT takes advantage of results like the Bubble concept, high-level architecture and several individual technology building blocks developed of the predecessor project DEWI[13] (Dependable Wireless Infrastructure, 2013–2017).

Many innovative applications of ICT in different aspects of society have been made possible by embedded computing resources together with innovative sensor and actuator systems locally into "things" (e.g., Smart Speakers). However, these systems often provide even more value when networked appropriately with other "things" in its vicinity. Arguably the biggest breakthrough however came by teaming up locality (embedded cyber-physical systems) with the close-to-infinite computation and storage resources of cloud computing. It is this combination which may make good on the promise of "real" smart things, providing disruptive innovations in almost all aspects of life (e.g., smart home, smart health and of course smart mobility).

In many contexts, networking smart things essentially means *wireless* connectivity, providing full mobility, and the means to establish connectivity anywhere, anytime and autonomously (no user intervention required). Wireless connectivity is therefore key enabler for the vast majority of such applications. However, wireless also brings challenges like varying quality of service, concerns regarding dependability (e.g., domains requiring certain levels of availability like industrial production lines; domains with for safety-relevant applications having regulations defining mandatory reliability). Aforementioned project DEWI focused exactly on this concern, defining the local DEWI bubble with dependability measures from physical all the way to application protocol layer, and appropriate access interfaces. Multiple demonstrators in automotive, aeronautics and rail domain clearly showed, that highly reliable wireless connectivity is achievable. Even so when sticking to tight energy budgets, which is another important factor for smart things (typically battery-powered and/or powered by energy harvesting from the environment).

DEWI indicated the way to dependable wireless connectivity. However, there was another major challenge to overcome in order to enable ICT everywhere: user acceptance. Users need to have trust in smart systems, and the services provided. Project SCOTT envisions "trustable things", which is addressed in several dimensions in parallel: from providing energy-efficient cyber security (confidentiality, integrity, and availability) both in the wireless communication bubble and beyond; to raising technology readiness level of the components of such smart things (including energy harvesting, efficiency and management). An important part of SCOTT is dedicated to analyze the way we humans establish trust in (computerized) systems. This includes

[13] http://www.dewiproject.eu/.

both security and other user-interaction based issues including contextual, organizational, and individual aspects (e.g., to inform users about the currently established level of system integrity and dependability).[14]

As envisioned by Mörtl et al. (2017), the trust framework developed and applied in SCOTT shall guide early design decisions and efforts to allow systems better trusted and accepted by its end users or acquirers. It will be applied and demonstrated by AVL and other industrial partners with 15 use cases in domains like automotive (connected vehicle, smart testing factory), health, production systems, aeronautics and train management.

SCOTT also suggests appropriate privacy labeling and trust indicating metrics for connected systems, to allow users to better judge the value and risks of using. This includes methodology to quantify privacy aspects like Miche et al. (2018), Fiaschetti et al. (2017). As an example, AVL cooperates with other partners on connected vehicles providing cloud-based services to travelers optimizing their mobility choices and monitoring health of their vehicles, including smart and predictive maintenance and software updates as required. Trust indicators and configurability supporting privacy labeling was part of the system design from the start. Other trust factors like system availability are addressed in technology building blocks balancing sustainable energy management and dependable connectivity with appropriate new protocol approaches like (Bernhard et al. 2018). Project SCOTT will display and report results in a multitude of demonstrators at its closure event in 2020; activities and progress are reported frequently via several channels.[15]

4 The IOT4CPS Project

IoT4CPS is a project run by a consortium of 16 Austrian scientific and industrial partners. It is partially funded by the "ICT of the Future" Program of the FFG (Austrian Research Promotion Agency) and the BMVIT (Austrian Ministry for Transport, Innovation and Technology). The lighthouse nature of the project is reflected in intention to not only deliver technical results, but also to provide guidance and inspiration for sustainable implementation of project findings in the post-project phase. Hence, the partners are also enhancing created impact by identifying and coordinating the interfaces to related ecosystems. One such example results from exploitation of the project's security nature, which makes it complementary to AutoDrive[16] an important lighthouse project for automated driving.

IoT4CPS focuses on development, production and operation of highly trustable components and applications for Connected and Autonomous Driving. The Smart Production is causally implicated into the project, as realization of the necessary components demands high levels of integration and information exchange along the

[14] https://scottproject.eu/home-3/trustability/.

[15] https://www.researchgate.net/project/SCOTT-Secure-Connected-Trustable-Things.

[16] https://autodrive-project.eu/.

life cycle. However, rather than diving deep into the full aspect range of Industry 4.0, the project's attention is turned towards the value chain of Autonomous Driving.

Many core technologies that form the set of complex mechanisms required for autonomous driving have already achieved levels of maturity needed for integration into commercially available vehicles. However, due to the rising complexity of the involved CPS, the utilization of the computing at the edge is also increasing. The dependability aspect of the proposed CPS solutions is a focal point of many discussions, as the safety is a critical factor in driving applications in general. Therefore, IoT4CPS is highly motivated to highlight the security and safety aspect of technologies capable of meeting the challenges related to automated and context-aware vehicles. The project aims to address safety and security aspects in a holistic approach along the value chain and the product life cycles. Hence, IoT4CPS is considering the entire technology stack from semi-conductors (sub-component level), through control systems (component level), to applications (system level). As the autonomous driving of the future is also very much likely to rely on information about specific aspects of external environment, the communication between the vehicle and infrastructure is also considered. The project partners are also estimating the potential development trajectories of various infrastructures and are modelling possible constructive interoperability of IoT platforms. Simultaneously, IoT4CPS addresses the product life cycle from development, via production to operational use.

The slowly emerging business models that utilize the technological advancements in connected and autonomous driving heavily rely on data collection, analysis and visualization. The key question is concerned with privacy, security and safety in terms of access to the raw data and to the results of the data processing. The concerns increase, if one considers that it seems nearly impossible to provide a true "end-to-end" IoT security solution from MCU level to Internet/Cloud. However, the efforts in terms of cryptography development are aiming at protecting the privacy of the users. IoT4CPS is seeking to implement efficient and secure cryptographic solutions, without introduction of undesired latency. The aim is to accommodate for the usage of dynamic number of IoT devices by enabling scalable and usable public key approaches. The analysis of communicated IoT data is proposed using security enabled tools that support identification of anomalies in a continuously changing IoT-enabled CPS systems.

The project is aiming to provide two types of industrial demonstrators. The autonomous driving demonstrator is going to be used for showcasing the implementation and usage of secure and safe IoT platforms and V2X communication. The Industry 4.0 demonstrator will showcase trustworthy radio connectivity solutions, life-cycle traceability solutions and security testing along the product lifecycle.

Figure 3 illustrates the project flow for the SCOTT and IOT4CPS projects. In both projects, the core value creation (trusted IOT technology bricks) is part of the ICT technology stack. At the same time, both projects are strongly driven by industrial use cases from different application domains. This organization very well highlights the different levels of value creation as presented in Sect. 2: the "*technology providers*" develop new technology bricks providing new capabilities. At the same time, "*end-to-end solution providers*", combining domain and ICT expertise, are required to

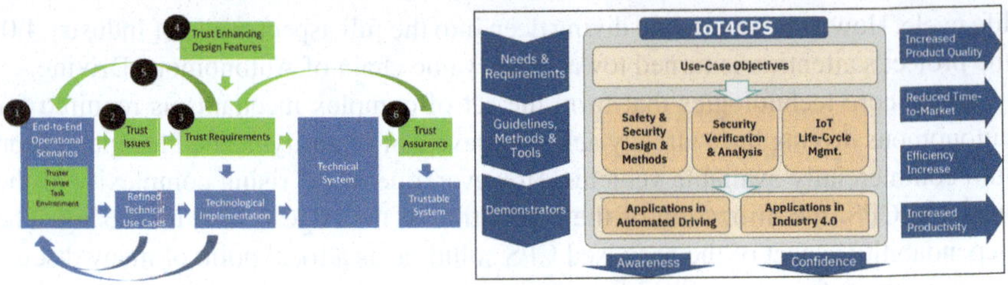

Fig. 3 Project flow SCOTT (left) and IOT4CPS (right)

finally integrate and tailor the technology bricks for a specific application domain. This cascaded value creation is probably one of the most important challenge for efficient opportunity identification, industrialization and exploitation of ICT solutions in industrial domains.

5 Summary

ICT is a game changer in the automotive domain. Through digitalization, the value creation is shifted toward more customized services around the vehicle, the user and the environment, leading to the emergence of new business models such as mobility-as-a-service. At the same time, ICT will require evolution in the value creation process. ICT is an individual journey for each institution. The European Commission has recognized the importance of this journey and is providing incentives and supports (e.g., through the PPPs and related roadmaps) to coordinate and accelerate its take-off, finally supporting the strengthening of the European industry toward digitalization. Still, a strong integration in the company strategy is required to go through this journey.

Acknowledgements This work has been supported in part by research from the SCOTT and IOT4CPS projects. SCOTT has received funding from the Electronic Component Systems for European Leadership Joint Undertaking under grant agreement No 737422. This Joint Undertaking receives support from the European Union's Horizon 2020 research and innovation programme and Austria, Spain, Finland, Ireland, Sweden, Germany, Poland, Portugal, Netherlands, Belgium, Norway. The IoT4CPS project is partially funded by the "ICT of the Future" Program of the FFG and the BMVIT.

References

Bernhard HP, Springer A, Priller P, Hörmann LB (2018) Energy balanced routing for latency minimized wireless sensor networks. In: 2018 14th IEEE international workshop on factory communication systems (WFCS), pp 1–9

Bosch J (2017) Speed, data, and ecosystems: excelling in a software-driven world. CRC Press

Bosch-Sijtsema P, Bosch J (2015) User involvement throughout the innovation process in high-tech industries. J Prod Innov Manag 32(5):793–807
Ebert C, Jones C (2009) Embedded software: facts, figures, and future. Computer 4:42–52
ERTRAC (2017a) Automated driving Roadmap, 7th version. Available at https://www.ertrac.org/index.php?page=ertrac-roadmap
ERTRAC (2017b) Urban mobility Roadmap, final version. Available at https://www.ertrac.org/index.php?page=ertrac-roadmap
ETPs (2017) "European roadmap electrification of road transport" jointly published by the three ETPs published in 2017. Available at https://www.ertrac.org/index.php?page=ertrac-roadmap
Fiaschetti A, Noll J, Azzoni P, Uribeetxeberria R (eds) (2017) Measurable and composable security, privacy, and dependability for cyberphysical systems: the SHIELD methodology. CRC Press
Hatler M, Gurganious D, Chi C (2012) Industrial wireless sensor networks. A market dynamics report. ON World
McKinsey and Company (2016) Monetizing car data—new service business opportunities to create new customer benefits. Advanced industries
Miche Y, Ren W, Oliver I, Holtmanns S, Lendasse A (2018) A framework for privacy quantification: measuring the impact of privacy techniques through mutual information, distance mapping, and machine learning. Cogn Comput:1–21
Mörtl P, Neuhuber N, Rosenberger M (2017) SCOTT public deliverable D28.1: "Foundations for Building Trusted Systems", V2.1, accessible via https://scottproject.eu/media/deliverables/

[illegible] (2018) [illegible]

Part II
Market and Trends

An Economic View on Electromobility in China

Peter Gresch and Jochen Siebert

Abstract The world's largest car market with currently about 24 million new registrations (total vehicle market 26 million vehicles) per year relies more and more on electric mobility. Thus, in principle, a (pre-)decision for the automotive industry worldwide regarding new drive concepts has been made. Also, by the end of 2017, more than 80% of the buses in China's megacities had already been converted to electric operation. By 2020, there will be more than 1.5 million electric cars and plug-in hybrids sold in China, with almost 54,500 electric cars sold in Germany in 2017, 117% more than the year before, yet Germany and Europe lag far behind. These are all irreversible signs that China is consistently advancing towards (E) automotive market leadership on the road to electrification of transport. Including topics such as disposal of lithium-ion cells, construction of charging infrastructure (target is one pillar per NEV per km in a few years), enough battery production capacity, stringent fuel consumption regulations and quotas for electric cars.

Keywords Transportation in China goes Electric

The world's largest car market with currently about 24 million new registrations (total vehicle market 26 million vehicles) per year relies more and more on electric mobility. Thus, in principle, a (pre-)decision for the automotive industry worldwide regarding new drive concepts has been made.

In China, a total of 470,000 electric cars and 105,000 plug-in hybrids were sold in 2017, 70% more than in the previous year. As a result, E-Vehicles market share in China rose from 1.4 to 2.4% in 2017. In addition there were about 300,000 eV commercial vehicles sold.

P. Gresch (✉)
PGUB Management Consultants GmbH, Darmstädter Straße 1, 64404 Bickenbach, Germany
e-mail: pg@pgub-consult.de

J. Siebert
JSC Automotive GmbH, Stafflenbergstraße 32, 70184 Stuttgart, Germany
e-mail: jsiebert@jscautomotive.com

J. Langheim (ed.), *Electronics Components and Systems for Automotive Applications*, Lecture Notes in Mobility,
https://doi.org/10.1007/978-3-030-14156-1_5

By 2020, there will be more than 1.5 million electric cars and plug-in hybrids sold in China, with almost 54,500 electric cars sold in Germany in 2017, 117% more than the year before, yet Germany and Europe lag far behind.

In addition to environmental aspects such as air pollution in cities or noise reduction, electric vehicles, that no longer need an internal combustion engine with mechanically complicated parts, offer China the opportunity to become the world market leader in the automotive industry as well. In addition, China can thus reduce the dependence on oil imports.

All vehicle manufacturers must achieve a fleet consumption of 5 L/100 km in 2020. If the consumption is higher, there are serious consequences, including a ban on the production of certain vehicles. All manufacturers will therefore pull out all stops to reach the consumption target. This ranges from more turbo and direct-injection over 48 V and plug-in hybrid to EVs. Plug-in hybrids with at least 100 km of electrical range and EVs count twice and are rated at 0 l/100 km.

The current requirement in China is that from 2019, automakers selling more than 30,000 cars a year in China will have to meet an electric quota for ten percent of their cars, by 2020 it will already be twelve percent.

If a manufacturer does not manage to produce enough cars with electric drive in 2019, he can make up for this with an increased production in 2020 or has to buy points from other companies. For this purpose, a “credit point trading” will be established. How much a point costs, is not fixed yet. The price is negotiated among the companies. So you have the choice between long-term negotiated contracts or go to the “spot market”. This is dangerous, because it might happen that there are not enough points to buy and the price per point rises sharply. Those who can prove not enough points are threatened by a government-mandated reduction in the production of conventional cars.

For each of these “quota cars” then a so-called “credit point” must be proven. Long-range electric cars receive several of these credits, at best up to six points. Plug-in hybrid vehicles (combustion engine and electric drive) get two points and fuel cell vehicles receive up to 5 points.

In the end, a manufacturer who has only electric cars and whose batteries guarantee a range of at least 350 km, would have to offer in 2019 only two percent of its vehicle sales electric. Most car manufacturers expect four points per electric vehicle.

For example, for an automaker such as Volkswagen, the largest manufacturer in China, this means that in 2019, with three million cars sold, around 75,000 electric models will have to be manufactured. There are no post-2020 targets, but we expect the quota to increase by 2% each year. This would be in line with the long-term plan, which was presented in 2017, but has not yet been passed into law.

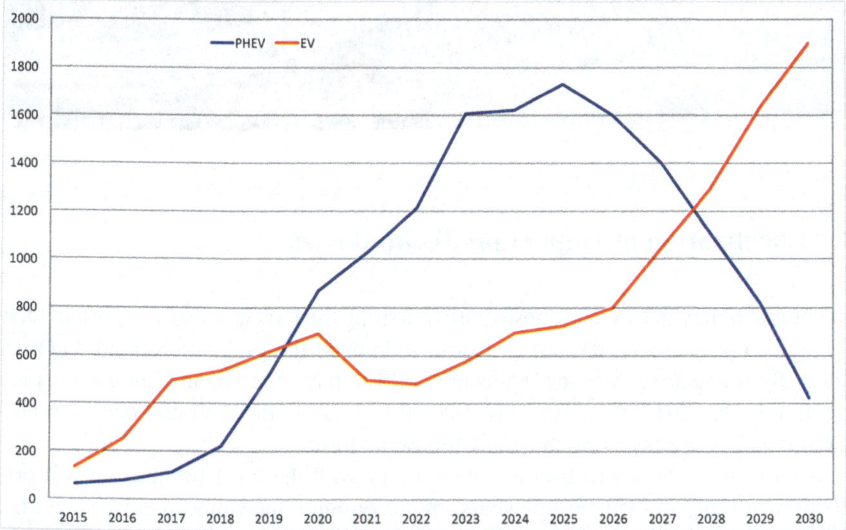

Forecast EV and plug-in hybrid sales China by 2030 (Source: JSC Automotive)

By 2020, a share of E-cars in the global new registrations of 2.5–6% is expected in 2025, it could then be 12–25%, 2030 even 25–40%. Of this, China accounts for at least one third. In the recent past, there has also been a great deal of dynamism between foreign and Chinese manufacturers as to who has entered into joint ventures with which Chinese NEV company or plans to do so, for example:

- Volkswagen is working with manufacturer Anhui Jianghuai Automobile (JAC)
- Ford announced that it would be merging with China's Zotye car group
- Renault-Nissan Group is planning a joint venture with Dongfeng Motor in 2019
- Honda Motor plans to build a battery factory in China in 2018
- China's car giant, Great Wall, which produced more than a million SUVs in 2016, but not a single electric car, bought 25% of the Yogomo company that builds cheap electric cars. Great Wall will also build electric cars with BMW.

1 Localization of Important Technologies

Economic policy in China is consistently pursuing the localization of important technologies. For example, the authorities have banned the usage of Korean batteries, allegedly for security reasons. Samsung and LG have been banned in the People's Republic since 2017. Consequently, like all local and other cell and battery manufacturers, they are investing in local Chinese factories.

As a result, China will become the country with the most lithium-ion cell production capacity in the next few years. In our opinion, the Chinese government has realized that it will be difficult for domestic OEMs to achieve high quality EVs. The pure subsidies have ultimately only created a market for substandard EVs, hence the new rules for NEV quotas are much stricter for use of high-tech.

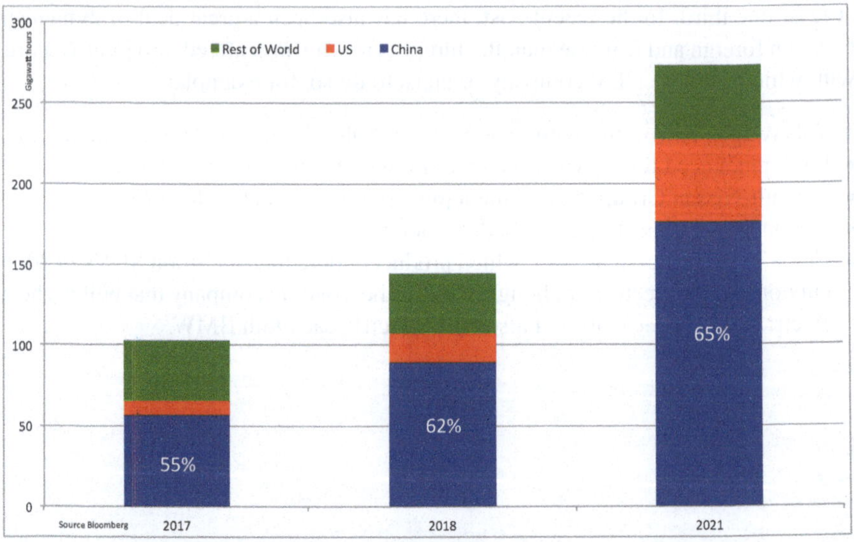

Global Li-ion capacities and China's share

China has also realized that even for EVs brands will be an outstanding feature for the attractiveness of a vehicle. Building a well-known consumer brand is much more difficult than succeeding in the capital goods business. That's why China is retreating to the "b-to-b" field, which is also easier to control. This also explains the new requirement that manufacturers of Li-ion batteries must have a capacity of at least 8 gigawatt-hours by 2020 in order to be approved as a supplier. China will roll out this market in a similar way to the market for solar panels a few years ago. Of course, the mass production of EV batteries and their very high level of subsidization will lead to significantly lower prices. However, a raw materials shortage of lithium is expected to occur from 2020 and will tend to drive up battery costs.

It is questionable to what extent the Chinese government, as in the case of the solar panel, is again breeding "zombie companies". Companies that are kept alive by government subsidies to save jobs, but do not make any operational profit and have no chance of improvement.

By far the largest battery manufacturer will become Contemporary Amperex Technology. The origin was Amperex Technology Ltd (ATL), which was founded in 1999 by Hong Kong based Robin Zeng. ATL was acquired by TDK in 2005 and Robin Zeng founded CATL in 2011 but remained TDK's top manager until 2016.

In 2015, ATL sold its 15% stake in CATL. The background to this was probably that subsidies from the Chinese state exclude the participation of foreign companies. CATL focuses on EV batteries and ATL on consumer batteries. ATL receives license fees from CATL. ATL and CATL are on the same premises in Ningde, Fujian and also share a research center there with about 1000 employees. Both companies work together otherwise, but there are no details.

CATL has offices in Sweden, Germany and France and also plans a factory in Europe. Shareholders are Chinese investors, as well as Foxconn from Taiwan, who assemble iPhones for Apple in China. Customers of CATL are amongst others BMW, Hyundai, PSA, BAIC, SAIC, GAC and Changan. CATL also bought 22% of Valmet in Finland in February 2017. The plan is to build and develop batteries and electric motors there.

CATL also has a JV with SAIC, which will produce batteries starting in 2018.

Their recent major move was the announcement some months ago to build a battery factory in East Germany.

Should our forecasts for the sales volumes of electric cars and plug-in hybrids prove correct, then significantly more battery capacity would be available than the market needs. Starting in 2020, a price war between the Chinese manufacturers is expected to set in and, similar to solar panels, China could try to capture the world market with fierce prices. This time, however, the affected countries would deliver an "epic battle" with China and not let this happen.

The Chinese government may then raise NEV quotas significantly faster to ensure survival for Chinese manufacturers. Therefore, one has to take the statements from the long-term White Paper literally and assume a quota of 40% for electric cars by 2030, and without multiple counting.

In this case, an annual battery capacity of 620 gigawatt-hours would have to be built by 2030 for a production of 10.3 million EVs and 2.8 million PHEVs.

Assuming 60 kWh for an EV and 7.5 kWh for a PHEV, we expect a required battery annual output of 618.1 GWh in 2030, which would give a total investment of about $ 80 billion from 2017 to 2030, including cell and cell capacity modules.

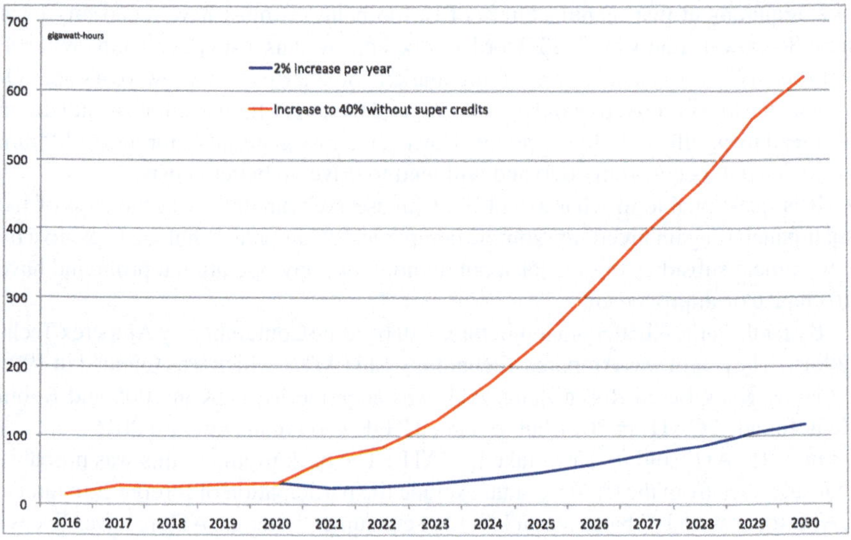

Capacity requirements for EV batteries in China 2016 to 2030

In Beijing, the maximum number of newly registered cars will continue to be reduced. If there were 150,000 new registrations in the city in 2017, there will only be 100,000 in 2018 and that will be at the expense of the burners alone. The number of electric vehicle registrations remains at the level of 2017 at 60,000, while the number of new registrations for burners drops from 90,000 to 40,000. In Shanghai, on the other hand, the number of registrations for burners is limited, but the number of registrations for electric cars is not.

2 Away from Subsidizing in a Few Years

Currently, there is the equivalent of up to € 7300 subsidy on the purchase of an electric vehicle

- However, only state-funded models are developed and built in China
- The Chinese government wants to boost the domestic automotive industry with the premium for electric vehicles.

However, as of 2021 there will be no more subsidy, as the program has proved too expensive (around US $ 80 total by the end of 2020) and not expedient, as most of the EVs are very simple and cheap. That is exactly what should change with the quota system.

By the end of 2017, more than 80% of the buses in China's megacities had already been converted to electric operation, with a total of about 43,000 large e-buses (over 10 m) and about 31,500 medium e-buses (6–10 m) built last year.

These are all irreversible signs that China is consistently advancing towards (E) automotive market leadership on the road to electrification of transport. Including topics such as disposal of lithium-ion cells, construction of charging infrastructure (target is one pillar per NEV per km in a few years), enough battery production capacity, stringent fuel consumption regulations and quotas for electric cars.

Part III
Electromobility

Silicon Based High Performance EV Batteries

Yohan Oudart

Abstract Nanomakers is a French start-up that manufactures at industrial scale four types of nanosilicon for high capacities Li-ion batteries for electric vehicles (EV):

- pure Si with particles size at 40 and 75 nm
- carbon-coated Si particles at 40 and 75 nm.

Those ultra fine particles are produced with narrow particle size distribution by laser pyrolysis. This process is reproducible, robust and the particles have low oxygen contents. The existing Li-ion technology uses for the anodes graphite with 350 mAh/g capacities. Nanomakers particles enable the manufacturing of high capacities Li-ion batteries anodes: 800 mAh/g along with a good capacity retention along cycling (>100 cycles). For silicon anodes, various formulations are studied but the most promising uses pre-composite in which the silicon is dispersed in a carbon matrix which is then partly decomposed by a thermal treatment. It is processed at a much lower temperature than graphitization processes. This micronic pre-composite can then be used as a high capacity raw material for anodes without changing anode preparation existing processes. The use of carbon-coated particles improves the anode performances during cycling, especially the capacity stability. This presentation compares the results obtained with various preparations and types of particles. The carbon coating has a positive impact on the batteries performances.

Keywords Silicon · Anode · Li-ion · Battery · Composite

1 Introduction

Lithium-ion (Li-ion) batteries represent a well-established technology for various applications and its use for electric vehicles (EV) is increasing exponentially due to the rise of climate change concern and pollution in cities. They are particularly

Y. Oudart (✉)
Nanomakers, 1 rue de Clairefontaine, 78120 Rambouillet, France
e-mail: youdart@nanomakers.fr

J. Langheim (ed.), *Electronics Components and Systems for Automotive Applications*, Lecture Notes in Mobility,
https://doi.org/10.1007/978-3-030-14156-1_6

suited for EV as they display both high energy storage and power, no memory effect (deterioration of the performances of batteries if they are not fully discharged before being recharged), and low auto-discharge.

However, the technology faces today the following issues braking its use in more demanding devices:

- Limited energy storage: All the application (portable electronics, EV) are looking for higher battery capacity. Indeed, with a battery autonomy ranging from 150 km (commuting cars) to 600 km (Tesla cars),[1] most EV autonomy are far below combustion engines equivalents. This lack of autonomy remains, alongside the price, the key barrier to electric car adoption.[2]
- Critical raw materials: graphite is an essential component in the production of the Lithium-ion batteries (anode component). The European Union Raw Materials Initiative reviews 27 critical raw materials,[3] among which natural graphite, commonly use in batteries.

Many materials have energy capacities higher than graphite, but silicon has the highest capacity with ten times fold the one of graphite (maximum of 4200 mAh/g for Si vs. 375 mAh/g for graphite).

This article describes how silicon will enable to improve EV batteries performances.

2 Nanomakers and Laser Pyrolysis

Nanomakers produces and commercializes silicon-based nanoparticles generated from silane precursor by a unique and patented technology developed at the CEA and of which the company is the world exclusive operator: Its principle lies in the interaction between gaseous reactants and a high power CO_2 laser beam under controlled atmosphere. The chemical bonds excitation causes reactant decomposition into atomic radicals which later gather to form particles whose growth is blocked by controlled quenching (see Fig. 1).

Laser pyrolysis has been developed in the MIT (USA) in 1979–1981 and then in CEA from 1985. Nanomakers is worldwide the only company using laser pyrolysis at industrial scale.

The advantages of laser pyrolysis are linked to the possibility of perfect control of the reaction parameters which lead to:

– Narrow particle size distribution

[1]https://www.statista.com/statistics/797331/electric-vehicle-battery-range/, consulted on December 2018.

[2]https://www.iea.org/publications/freepublications/publication/GlobalEVOutlook2017.pdf, consulted on December 2018.

[3]https://eur-lex.europa.eu/legal-content/EN/TXT/?uri=CELEX:52017DC0490, consulted on December 2018.

Fig. 1 Laser pyrolysis process with CO_2 laser that encounter reactant gas

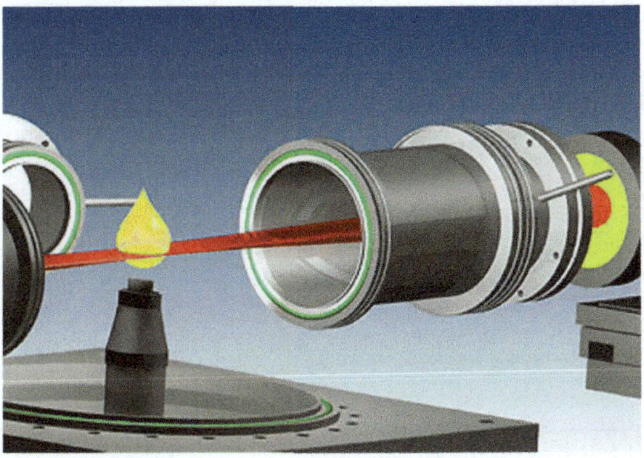

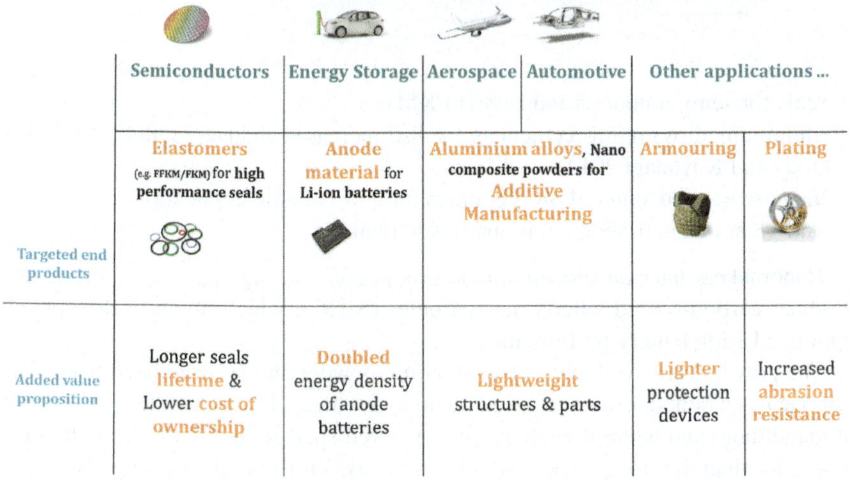

Fig. 2 Nanomakers nano silicon carbide and nano silicon main application

- Very low metallic impurities
- Low oxygen content (<2%)
- Good reproducibility
- Tunability of size (30–100 nm), surface, coating, …
- Industrial capacity (up to 20 t/year per line).

Nanomakers is therefore fully able to provide best-in-class nanopowders at industrial scale and controlled costs. The quality of its products has been recognized both by academics (Shimoda and Koyanagi 2014) and industrials (Eck Industries) in that respect.

Silicon carbide nanopowders have been first produced to be used for mechanical reinforcement of materials but its real interest lies in its multiple properties (See Fig. 2):

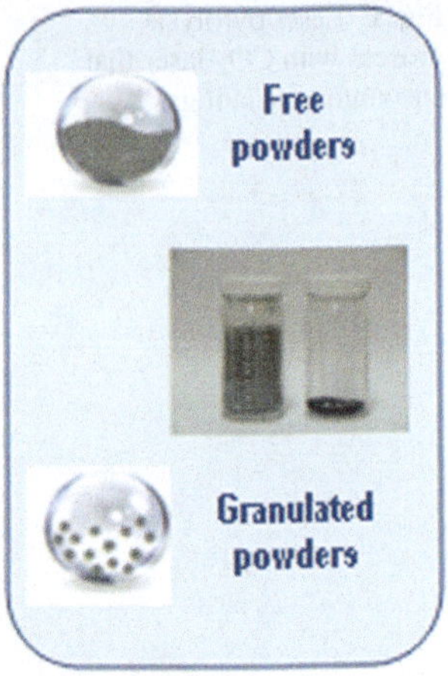

Fig. 3 Comparative volume of the same weight of granulated (right bottle) and free powder (left bottle)

- Seals for semi-conductor industry (FFKM)
- Aluminum alloys reinforcement as 1% SiC increases yield strength by 95% (Shimoda and Koyanagi 2014)
- Naonostructured sintered SiC ceramics for anti-ballistic application
- Anti-wear and anti-abrasion properties in plating.

Nanomakers later developed silicon production and then patented a brand-new product: carbon-coated silicon nanoparticles (SiΩC), which allow to disruptively optimize Li-ion battery performances.

The powders are available as free standing powder and as granulated from. This last one has a bulk density decreased by up to ten times, lower dustiness (2–3 orders of magnitude) and better flowability. It can be redispersed in order to have the same properties than the free powder. No additive is used for granulation (Fig. 3).

3 Why Silicon Will Be Used in Batteries?

Numerous elements can be used in anodes batteries with potential higher specific capacities than graphite. Silicon exhibit the highest specific capacity compared to other materials that often also suffer from poor cyclability or lack of large availability (Ge, Ga, In) (Fig. 4).

SiO_x has paved the way to silicon use in EV batteries to increase their capacity by 20% in the anodes. However further improvements seem to be limited and silicon metal will take over from SiO_x to increase anode capacities by 50–100% in the coming years.

Silicon metal suffers from two main issues during cycling:

- when too large, particles are pulverized during battery cycles and silicon losses its electric connexion to the anode. Characterisation studies succeeded in defining

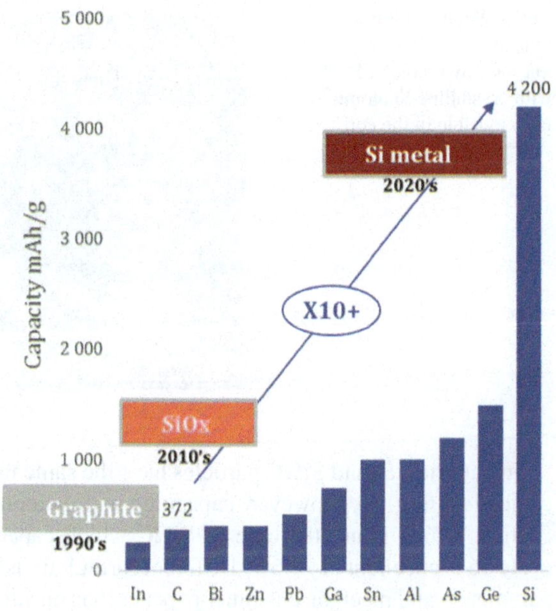

Fig. 4 Maximal element specific capacities in Li-ion anodes

a maximal size of 150 nm under which pulverization effect is limited (Liu et al. 2012).

– when too small, particles are oxidized during cycling due to the reaction with electrolyte. This reaction increases along with the particles surface increase.

Depending on the system, the good size compromise ranges from 30 to 80 nm with a frequent optimum found around 40–50 nm.

Silicon metal is not used pure as its very high initial capacity fade rapidly. It has been first diluted in graphite and is now incorporated in a carbon composite. It is made by mixing the nanoparticles with a carbon matrix which is then pyrolyzed to change its structure and enhance interactions with silicon. This product is then crushed to 10–40 μm, a size similar to commercial graphite for Li-ion batteries. In this form it can be readily used without changing the anode manufacturing tools. The anode active materials current targets are 600–1000 mAh/g which correspond to 10–35% wt Si depending on the material co-products which also have lithium storage capacity (carbon additives, pyrolyzed carboned matrix, graphite, binder,…).

4 Why Coated Silicon by Laser Pyrolysis?

For SiΩC, the production of the silicon crystalline core and the carbon introduction are made in one single stage during laser pyrolysis. This product surface is amorphous as shown on the Fig. 5.

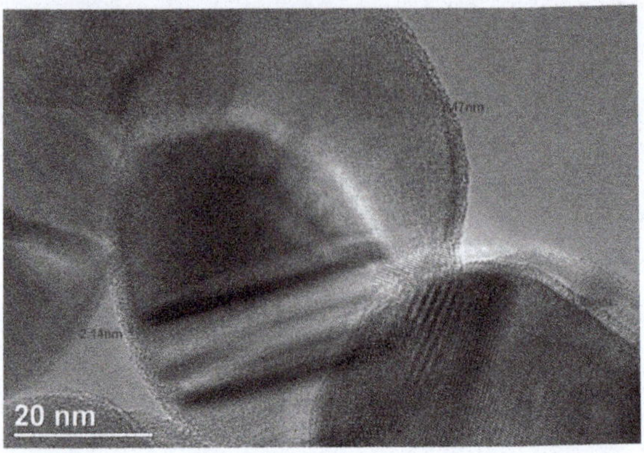

Fig. 5 High resolution transmission electronic microscopy image of SiΩC, with crystalline Si atomic layers visible in the core

In batteries, Si and SiΩC particles have the same initial capacities, even after few cycles (see Fig. 6). However, capacities with silicon only decrease slowly during cycling, which is not the case of SiΩC which capacity retention from cycle 2 to cycle 86 is excellent at 98.6%. Different formulations can be made to optimize other parameters and meet the customer expectations on various criteria (specific capacity, initial coulombic efficiency, capacity retention after 100 or 500 cycles, swelling ratio,…).

This new kind of nanoparticle displays numerous advantages over mere silicon nanoparticles for instance:

- Silicon protection from the electrolyte and oxygen;
- Stability enhancement of the Solid Electrolyte Interface (SEI);

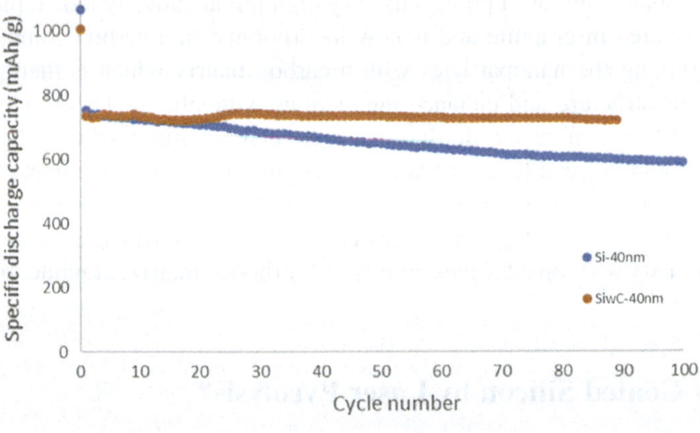

Fig. 6 Compared performances of coated and non-coated silicon (half cell, 80% composite, 10% SuperP, 10% CMC; electrolyte EC:DEC 1:1 +2% wt VC +10% wt FEC +1 M LiPF6, lithiation CCCV till 10 mV, delithiation CC 1 V, C rate: C/5 (C/20 on first cycle), Helios project result, to be published)

- Higher affinity between silicon nanoparticles, graphite and other binders;
- Uniform distribution of silicon nanoparticles in the carbon matrix;
- Increase in ionic and electronic conductivities thanks to the carbon layer.

5 Conclusion and Perspectives

Nanomakers produces and manufactures high quality silicon based nanoparticules. They are unique not only due to their industrial production process (laser pyrolysis) but mostly due to their properties (narrow particle size distribution, high purity, reproducibility and customization in terms of size and surface properties). Nanomakers historical products is silicon carbide for various materials mechanical reinforcement (elastomers, metals,...) as additive or nanostructured SiC parts. The second product is silicon nanoparticles that can be carbon coated. Their main use is Li-ion batteries as it has a specific energy storage capacity ten times higher than actual graphite (375 mAh/g). On short term anode active materials with metallic silicon will reach 600–1000 mAh/g. Silicon is incorporated in a composite made of pyrolyzed mix of carbon and silicon. There is a consensus for the use of 30–80 nm particles which is the good compromise between low size required to avoid cracking during cycling and low exposed surface to minimize oxidation by electrolyte. SiΩC (Carbon coated particles) exhibits better cyclability than the non-coated particles due to oxidation protection and better contact with the carbon matrix.

References

Liu XH, Zhong L, Huang S, Mao SX, Zhu T, Huang JY (2012) Size-dependent fracture of silicon nanoparticles during lithiation. ACS Nano 6:1522–1531

Shimoda K, Koyanagi T (2014) Surface properties and dispersion behaviours of SiC nanopowders. Colloids Surf. A 463:93–100

Predictive Electronics for Improved EV Battery Tray Monitoring

Lionel Bitauld and Joseph Bosnjak

Abstract Batteries (mostly Li-ion) for electric and hybrid vehicles are integrated in a metal structure (battery compartment) associated with the chassis. A dedicated electronic (called BMS, Battery Management System), controls and balances the voltage between the different elements of the battery. However it appears this monitoring does not cover all the failures mode of the battery or at least it is possible to better anticipate their occurrence. Glycol water mixture coming from the cooling system, abnormal concentration of methane, overheating, shocks the battery compartment would have absorbed and presence of smokes are all relevant of a battery failure. As a supplier of metal envelops for the battery, DURA has developed a complete system to detect those threats, in particular the liquid and the smoke detection systems. The set of sensors is placed in the battery compartment, while the ECU which manages them and integrates the various strategies and also communicates the resulting information to the vehicle is located outside this compartment.

1 Background

Batteries (generally Li-ion) which store the energy required for the propulsion of the electric and hybrid vehicles are integrated into a metal structure (battery compartment) associated with the chassis.

These batteries are made of a multitude of elementary cells (3.6 V nominal voltage) arranged in series and in parallel in order to reach the required output voltage and total needed energy.

A dedicated electronics (called BMS, Battery Management System), controls and balances the voltage between those cells. It also manages the charge/discharge and evaluates the remaining amount of energy available in the battery.

L. Bitauld (✉)
DURA Automotive Systems, Bievres, France
e-mail: bitauld.l@duraauto.com

J. Bosnjak
Solairion, Jouars-Pontchartrain, France

J. Langheim (ed.), *Electronics Components and Systems for Automotive Applications*, Lecture Notes in Mobility,
https://doi.org/10.1007/978-3-030-14156-1_7

It appears however that this monitoring does not cover all the failure modes of the battery or at least their detection can be improved.

We can thus quote in particular the presence of water in the battery compartment, an abnormal concentration of methane, an abnormal temperature of the envelope, monitoring of the shocks the battery compartment would have suffered and finally smoke detections.

All those defaults are the sign of a battery concern.

2 Intended Purpose

The battery compartment isn't only bulky (one to two m3) is also a complex component. Generally made of aluminium for mass reasons, it has to respond to numerous mechanical constraints (stiffness, mass, tightness) which make its design and assembly during production (control of welds, sealing) tricky. DURA, through its Structural Systems division, provides such products to its automotive customers (Fig. 1).

Thanks to its embedded electronics skills, DURA can also design and integrate electronics in the battery compartment to address the problem described in the previous chapter.

This has driven to a dedicated electronic architecture (number of sensors, type of communication, data management and implementation of algorithms) and a mechanical architecture (location of sensors, design of envelopes for sensors, choice of connectivity) (Fig. 2).

Fig. 1 Example of battery tray

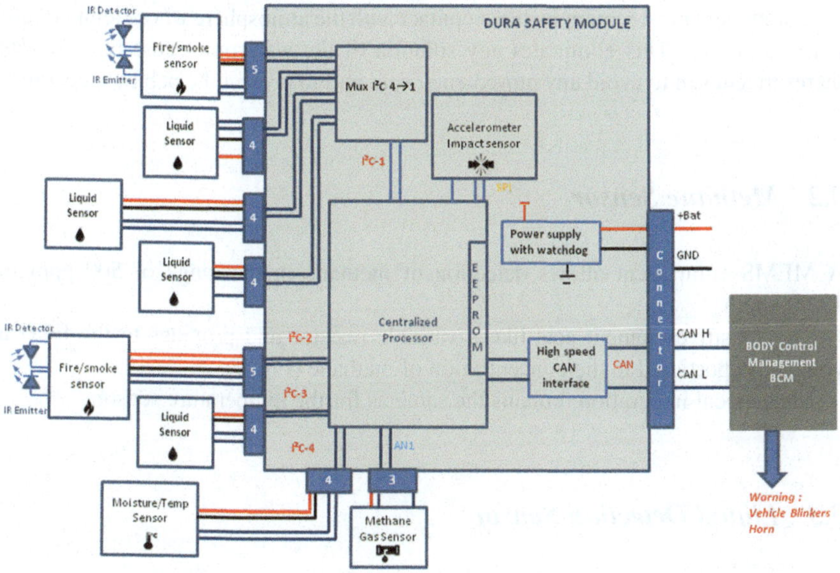

Fig. 2 System architecture

3 Results and Originality of the Solution

3.1 Temperature Sensor

Design is based on a capacitive digital sensor which also performs the moisture measurement needed for the methane sensor compensation.

The ECU then acquires data which are filtered and compared with a threshold (maximum operating temperature of the batteries under normal conditions) and communicates the information to the vehicle (Fig. 3).

Fig. 3 Temperature sensor pcb

Component must remain in direct contact with the atmosphere whose temperature is to be known. This eliminates any solution of the waterproof case type. It was therefore chosen to avoid any closed enclosure and to protect the pcb by a coating.

3.2 Methane Sensor

A MEMS component allows detection of methane in the range of 500 ppm to 10,000 ppm.

The sensitive element acts like a variable resistor and provides to the ECU a voltage proportional to the concentration of methane (Fig. 4).

Mechanical integration remains the same as for the temperature sensor.

3.3 Liquid Detection Sensor

Design is based on a capacitor which consists of 2 copper tracks on a pcb and whose dielectric is either made of air (no leak) or made of liquid (when leak occurs). As relative permeability of liquid (70–80) is higher than the one of air, capacitance value will increase.

A dedicated driver operating in differential mode to improve sensitivity measures the capacitance (Fig. 5).

Fig. 4 Methane sensor pcb

Fig. 5 Liquid detection pcb

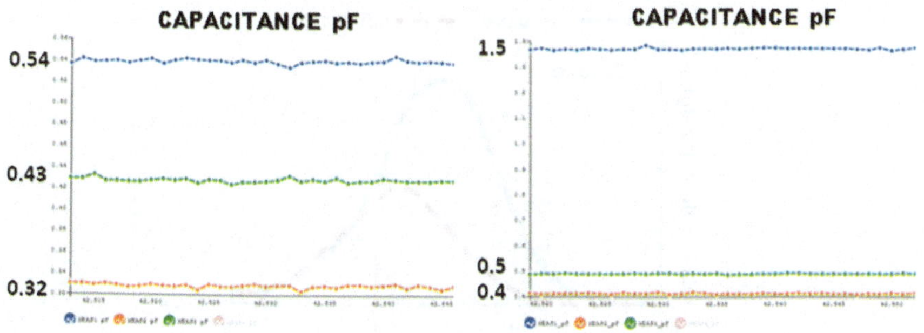

Fig. 6 Capacitance change

Two identical capacitances are used to perform detection of malfunction (safety issue). Those capacitances are implemented on a pcb which is horizontally installed in the bottom of the battery compartment.

A few amount of liquid induces an important change of the capacitances value (change from 0.5 to 1.5 pF) (Fig. 6).

3.4 Smoke Detection Sensor

A photo detector measures the amount of light retro diffused by the smoke particles when they are illuminated by a light source (generally infrared LED).

As explained below two light sources of different wavelengths are used. The photo detector is placed next to them.

In the absence of smoke, the intensities received by the photodetector are identical.

In the presence of smoke particles much smaller than those present in the air, the intensities received differ. According to the principle of Rayleigh they are proportional to the fourth power of the wavelength of the emitted signal. By using an infrared source and a blue source, the ratio of the two retro-diffused signals can reach 16 (see Fig. 7).

More specifically, when smoke is present ratio between blue and IR light is greater than one. It will remain higher than one as long as smoke will be present.

In case of steam for example, it will remain cqual to one (Fig. 8).

3.5 Shocks Detection Sensor

Shocks are measured by the mean of an accelerometer located in the ECU on one wall of the battery tray.

Detection principle is very similar to the one used for airbag application.

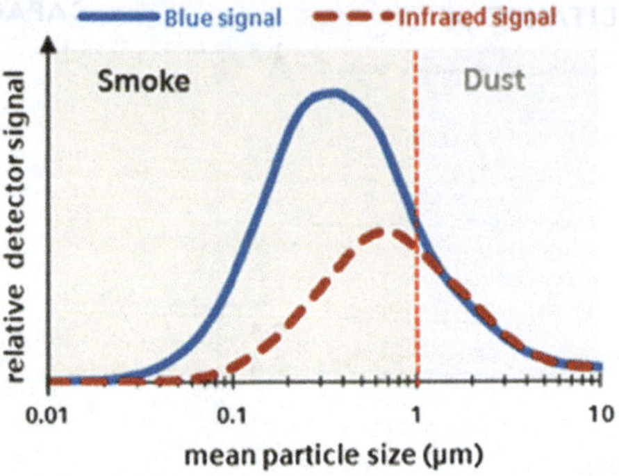

Fig. 7 Blue and IR signals relative intensity

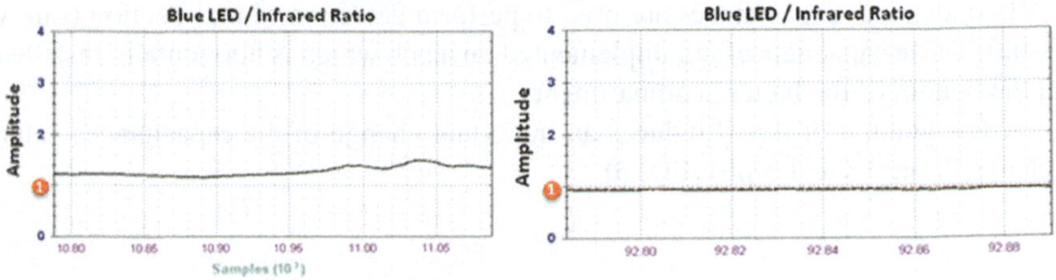

Fig. 8 Ratio between blue and IR signals for smoke (left) and steam (right)

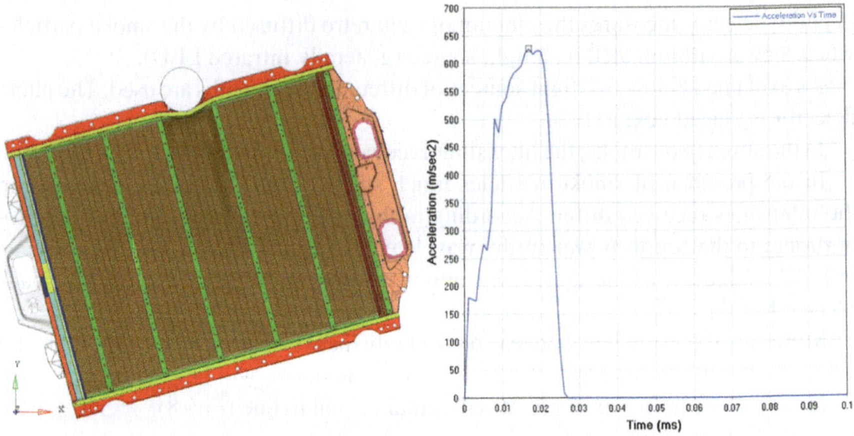

Fig. 9 Simulation for acceleration when shock occurs

Bumps are detected via the calculation of the Jerk signal (derivative of acceleration). Jerk is then compared to a jig made of several thresholds (Fig. 9).

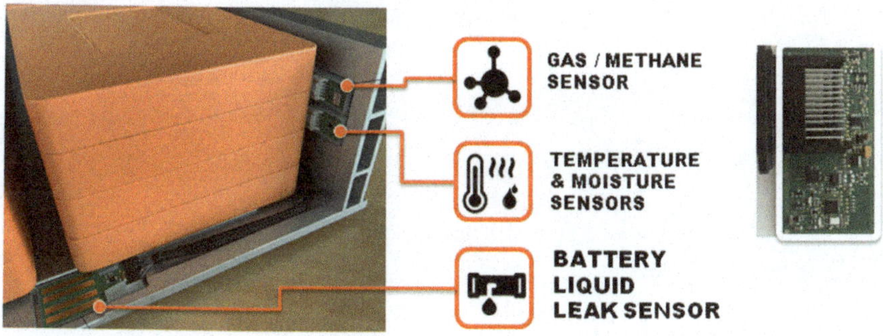

Fig. 10 System overview in a battery tray

4 Conclusions

BMS monitors battery parameters and manages the performance of the battery system.

Unfortunately monitoring battery variables may not be sufficient from a safety perspective.

DURA battery tray safety module monitors and analyzes critical environmental parameters in the physical vicinity of the battery system.

This provides an independent and reliable way of assessing the health of the EV battery system and to mitigate any safety risk.

DURA's approach helps to provide a truly redundant complete system (Fig. 10).

References

Aspirating smoke detection, White Paper I 09 June 2015 by Siemens
Capacitive-based liquid level sensing sensor reference design by Texas Instruments

Battery Management System: From Safe Architecture Definition to System Simulation with Embedded Software

Xavier Fornari

Abstract Batteries development has changed through the past years with the emergence of electric vehicles. The sustainability of the battery must be ensured along with its safety as they deliver high voltage and the technology used is more difficult to control. Moreover, these batteries may be used in full autonomous vehicles, with no direct human control. In this paper we will show how to set up a complete workflow from the safety requirements down to the software implementation while fulfilling the ISO 26262 objectives. We also show the benefits of using a model-based approach and the gains that can be obtained on some testing or reviews activities. Finally, we complete the flow with the integration within a full-virtual system, which allows for implementing the hardware design, as well as the integration testing.

Keywords Battery management system · Safety analyses · Model-based system engineering · Certified code generation · Virtual system simulation · Reduced-order model · ISO26262

1 Introduction

Batteries are keys in today's technologies. There are present almost everywhere from the smallest objects to the biggest ones. The mobility area is a domain that is demanding for new batteries, as they will be used in Hybrid, Plug-in Hybrid and full electrical vehicles. Such vehicles are not only cars but will be aircrafts including drones.

X. Fornari (✉)
ANSYS, Montigny-le-Bretonneux, France
e-mail: xavier.fornari@ansys.com
URL: https://www.ansys.com
URL: https://www.linkedin.com/in/xavierfornari/

Esterel Technologies—A Wholly-Owned Subsidiary of ANSYS, Inc., Montigny-le-Bretonneux, France

J. Langheim (ed.), *Electronics Components and Systems for Automotive Applications*, Lecture Notes in Mobility,
https://doi.org/10.1007/978-3-030-14156-1_8

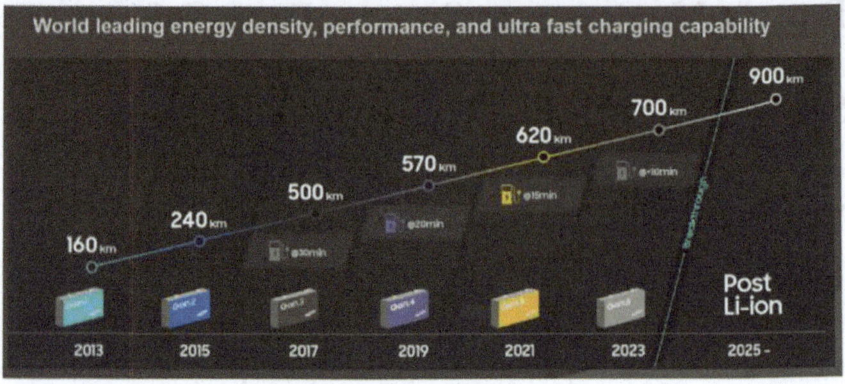

Fig. 1 Batteries evolution (*source* Lee 2018)

Mobility challenges require an increase of two times in battery energy density and a 50% reduction cost/kWh. Technology is moving fast in this domain, as suppliers are challenged with providing batteries that go longer distances and have a longer life time (Fig. 1).

Unfortunately, batteries present a tough engineering challenge. They combine:

- Multi-physics: electrochemistry, thermal conduction, physical structure;
- Complex use: low temperature leads to less power, high temperature has an impact on the life-time, operation generates heat, …;
- Crash, vibrations, situations that can leads to thermal runaway, then explosion;
- Charging/Discharge/Recuperation process has a strong impact on life time.

New generations of batteries for cars and aircrafts provide more that 400 V, which is now dangerous with the delivered current.

Therefore, not only do these objects become more and more complex to design, but they are increasingly difficult to control and it is mandatory that safety is ensured. In 2010 the US Department of Energy (DOE) launched the Computer Aided Engineering for Electric Drive Vehicle Batteries (CAEBAT).[1] DOE found there is a lot of trial and error in battery design (CHARGED magazine 2012).

In this article we present how it is possible to develop such complex batteries while ensuring their safety.

Section 2 of this paper is about architecture and functional safety. To guarantee the safety of the battery, all possible risks must be analyzed to derive the safety goals and the proper architecture. Unfortunately, this activity is often done "manually", using various disconnected methods, and there is no real interaction between the safety architecture and the design architecture. We show that software tools may help to ensure a consistent analysis of the architecture while keeping a link between safety engineers and design engineers.

---

[1] Computer-Aided Engineering for Electric-Drive Vehicle Batteries (CAEBAT). https://www.nrel.gov/transportation/caebat.html.

Section 3 focuses on the battery hardware, which includes the design of the cells and the pack. This is indeed a multi-physics, multi-scale and multi-dimensional problem. Simulation is now able to tackle all of these point of views in a comprehensive way, allowing for performant battery design.

The next step is the control software of the battery. This software becomes more and more complex due to the nature of the battery, and in addition, its criticality becomes higher. The reasons are that the battery object is more dangerous, and care must be taken with it. Also, these batteries will be used more and more frequently in autonomous vehicle. Developing software under such circumstances is quite costly, as great levels of confidence must be achieved. Section 4 shows how model-based development, connected with safety as described in Sect. 2, can provide a sound and economic answer.

Finally, Sect. 5 brings everything together, the hardware and the software. Complete virtual simulation of the battery with its control is brought into a realistic environment to perform architecture/performance analyses at various stages of the development. Ultimately, this allows for functional verification and validation to ensure that the safety goals are achieved.

2 Architecture and Functional Safety

To fulfill the safety requirement of a battery a structured approach is required. Following the safety lifecycle for the ISO 26262 standard (see Fig. 2), the first steps are the analyses for hazards and the definition of the functional safety concept, before moving to the hardware and software part. The first difficulty is to perform multiple analysis methods in a consistent, thorough manner: hazard analysis, failure mode and effects analysis (FMEA), fault tree analysis (FTA), controllability analysis, while including change and configuration management. The second difficulty is that the safety engineers are not necessarily connected to system engineers. Both teams may start working on different concept architectures, while there is a strong need for communication to address the safety goals.

A consistent process would address these two difficulties by allowing one to achieve all the analyses within a single framework and to exchange architectural information between teams.

Using a model-based approach to safety resolves these goals. The model of the architecture is at the center of the analysis activities as shown in Fig. 3. All the activities are correlated as they are done in a single framework, utilizing the same model. As such, any change to an item of the model has a direct impact on associated analyses, or derived analyses. A model-based approach therefore ensures unrivalled levels of consistency, traceability and efficiency. The output of that process is an architecture that includes both the hardware part and the software parts. The architecture may define redundancy of components or specific information flows to ensure the safety goals.

3 Battery Design

This section covers the physical design of a battery hardware. Battery design is a complex task, involving kneading (electrode), die internal flow, coating, calendaring and compression of porous electrodes, drying of electrolyte, sharp cutting of foil, electrolyte filling. But it is also about chemistry, electricity and flow–thermal–structural physics inside batteries that operate in a complex interplay that must be strictly controlled within narrow margins to ensure long battery life and low cost. The slightest deviations from these margins can lead to dramatically shortened battery life—and to serious safety hazards.

One of the major concerns in the development of lithium-ion battery packs for electric vehicles is thermal management. The temperature of all cells must be strictly maintained within a few degrees Celsius across the entire pack. Cells are vulnerable to overheating from rapid discharging, overcharging or excessive ambient heating. Such overheating and uneven temperature distribution can lead to rapid cell degradation and shorten battery life. In extreme cases, thermal runaway may occur in a cell as heat builds up uncontrollably, possibly causing catastrophic destruction, such as fire and explosion.

Simulation is the mean to design the batteries and to consider these effects. It requires advance technics as the problem is:

- **Multi-dimensional**: the system is composed for sub-systems and components with a mix of modeling capabilities ranging from 0-D (like control logic) to 4-D (time varying 3-D flows for instance);
- **Multi-physical**: obviously, a battery is composed of different physics, and one wants to study thermal effects on structure, electrochemistry phenomena within cells or cooling rates;

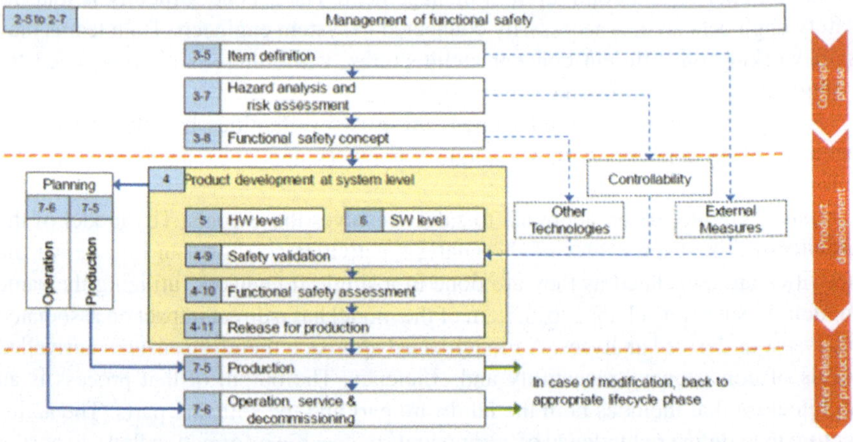

Fig. 2 ISO 26262 safety lifecycle

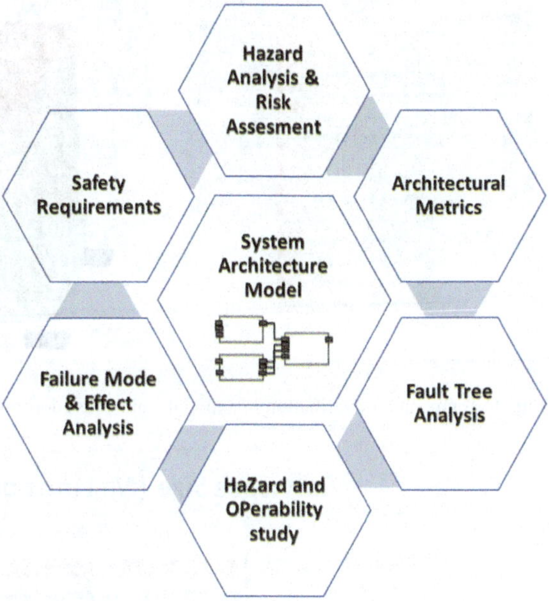

Fig. 3 Model-based approach to safety using medini: central architecture model, extended with analysis-related properties

- **Multi-scale**: electrochemical effects occur at nanoscale, while the battery controller works at the pack level. The capability to span these multiple levels is critical in evaluating the many interrelated aspects of battery development, particularly issues related to thermal management.

Physics-based electrochemistry models can be used to more readily investigate the impact on battery performance of design parameters including battery geometry properties and, most importantly, temperature. The most widely recognized physics-based model originally was proposed by Professor John Newman from the University of California Berkeley. For higher-level analyses and simulation, an equivalent circuit model can be used. Such equivalent circuit can be derived from existing battery cells using parameter identification (Hu et al. 2013). Other technics such as Reduced-Order Models (ROMs) allow for extracting the information from the 3-D simulation. Such models can run in a few seconds instead of hours of CFD computation (Hu et al. 2011). System-level design engineers can then concentrate on parameters and easily make updates with a few clicks (Fig. 4).

4 Software Design

Software is present in every industry, but in automotive, it appears to not only have a large part (it is estimated that more than 50% of the electronics components will be software), but also to be huge, as Fig. 5 shows. Software is everywhere: engine con-

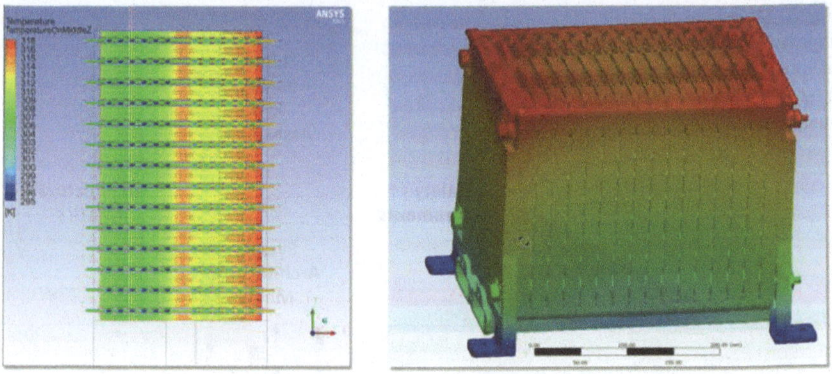

Fig. 4 Temperature distribution in a module and stress deformation

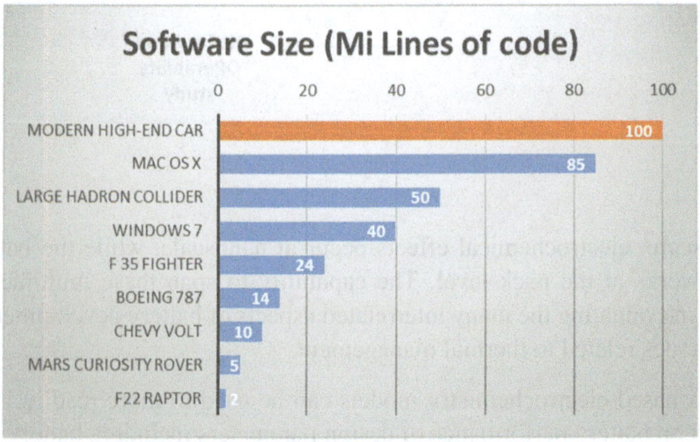

Fig. 5 Relative software sizes

trol, ABS, power train, battery, climate control; lightning, and it becomes critical due to the connected autonomous car trend.[2] Autonomy is viral. For instance, software for a braking system or an electronic steering power system have their criticality raised to ASIL D, because there may be no driver in future cars that could take an emergency action if needed.

For the specific case of batteries, the critical aspects are on one hand economic and on the other hand related to safety. The objective is to have efficient batteries that last, delivering the right power when needed, while controlling use to avoid thermal effects that may lead to explosion. New generations of batteries are in the range of 400 V rather than 12 V. There is a need to properly manage the State-of-Health (SOH) and the State-of-Charge (SOC), in order to maintain a Safe Operation Area

[2]Driving towards autonomous cars and ADAs—the future is now, blog, https://www.ansys.com/blog/autonomous-cars-adas/.

(SOA). The software in charge of the control is the Battery Management System. It combines complex algorithms that guarantee the proper behavior of the battery with respect to its estimated state and it has to guarantee the safety of the battery system. This means two things:

1. The system and its software are developed following the safety analysis, as described in Sect. 2. As example, an emergency disconnect sub-system is required;
2. The software itself must be develop in a way that full confidence can be obtained.

To achieve the proper level of confidence, the automotive industry developed the ISO 26262 standard. The standard describes a life cycle for the software and electronics parts, with different phases. But the process is particularly costly due to the number of reviews, verifications and traceability analyses that have to be performed (Fig. 6). In particular, for ASIL D level, the highest level of safety, structural coverage is required which is particularly long and costly to fully achieve. In addition, any change has a strong impact on the whole chain of activities. Overall increase of costs from one ASIL level to the next ASIL level is given (see Table 1).

Use of a model-based development environment and certified code generators is certainly the state-of-the-art practice to achieve the expected high level of confidence while maintaining the time-to-market and the costs. SCADE (Fig. 7) is an environment based on the Scade language specifically designed to model and program high-integrity applications. The semantics of the language ensure that the designed model is not ambiguous, which is key for communication between stakeholders, and the provided constructs fulfill safe embedded code constraints (no pointers, bounded

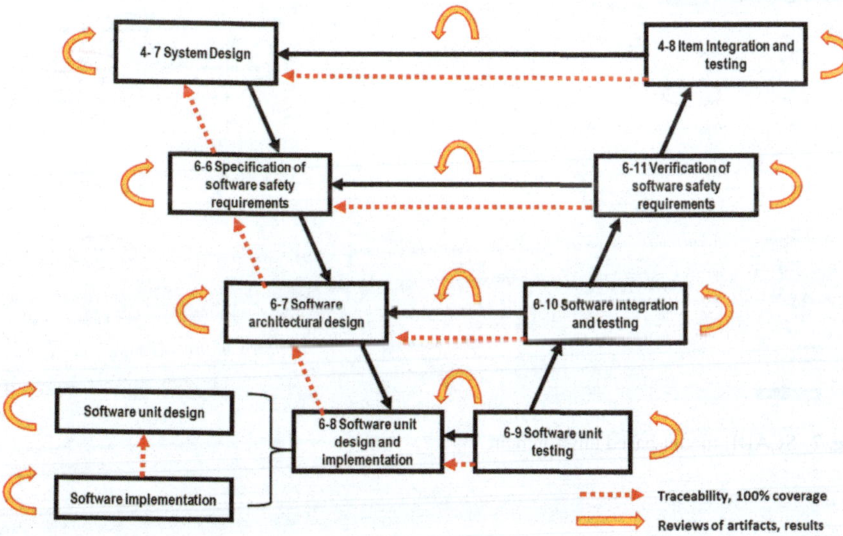

Fig. 6 ISO 26262 software activities

Table 1 Relative increase of certification activities costs

	ASIL A (%)	ASIL B (%)	ASIL C (%)	ASIL D (%)
Plans	10	38	24	14
Architecture	10	30	17	25
Design&Code	10	50	20	15
V&V	11	38	27	151
Total	9	38	21	71

loops, …). The model is actually the specification of the software: all activities must be done at that level. The certified code generator produces C code that does not need to be verified and fully implements its specification. This is ensured by the fact that the code generator *itself* has been developed following the ISO 26262 recommendations,[3] and this has been assessed by various authorities, including TÜV, FAA, EASA, and others. The environment also offers traceability capability with requirements and tests, and synchronization with the software architecture. This later point is fundamental as it is the link between software design and safety analyses.

This model-based process from architecture/safety-design-code generation generates interesting gains in terms of development costs (time/resources) (see Table 2), but moreover it gives strong confidence in the final code.

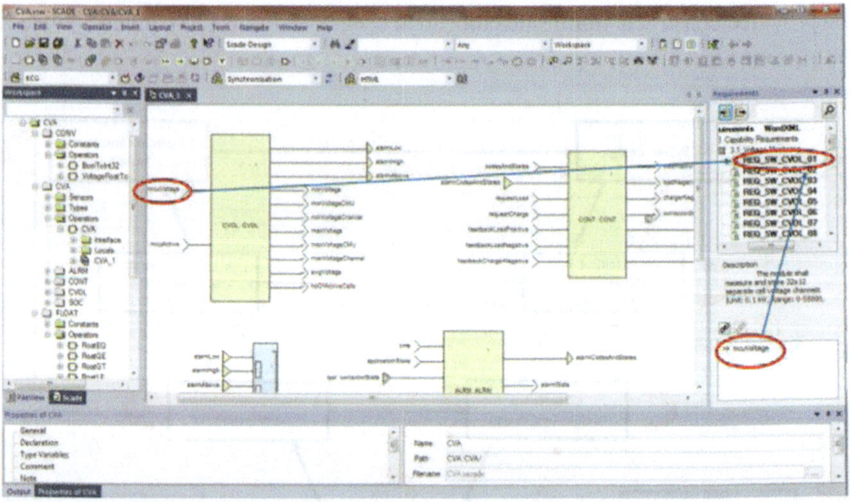

Fig. 7 SCADE model-based environment

[3] It also follows those of DO-178C for Avionics, EN-50128 for Railways and Transportation and IEC 61508 for Industry.

Table 2 Full model-based software design certified process gains

	No Cert (%)	ASIL A (%)	ASIL B (%)	ASIL C (%)	ASIL D (%)
Plans	0	70	75	80	80
Architecture	10	10	20	20	20
Design&Code	53	53	48	48	50
V&V	35	48	59	59	58
Total	36	46	51	52	54

Another key point to consider is the openness of such environments so that they can seamless be integrated in existing flows. Openness can range from support of existing standards, like AUTOSAR,[4] to coupling with existing tools via APIs,[5] or interacting with these by the means of gateways, like Simulink model importation or S-function export. It is a mean to protect existing assets while moving to new practices.

5 System Simulation

We considered the development of the hardware part of a battery, using fine grain simulation, and the development of the software part, taking into account the safety of the system and the need to support an ISO 26262-based process.

The next step is to bring everything together and to go for system simulation. Here system simulation is the simulation of the combination of different components, mixing various physics and eventually control software. We do not consider system simulation as it may exist using SysML behavioral diagrams.

The Fig. 8 depicts a complete virtual simulation that combines:

- The battery pack and its software controller;
- The electric drive and its software controller;
- The cruise-controller application fed with a mission profile;
- A model of the car dynamics, including body and brake system.

This kind of simulation is now available and it possible to bring together components using environments based on Modelica[6] or by combining FMU (Functional Mock-up Units[7]). Components can be described using various models: C/C++, Modelica, VHDL-AMS at different levels of accuracy. A key technology here is the

[4] AUTOSAR—https://www.autosar.org/.

[5] API: Application Programming Interface. A API can be provided as Python scripting access to data.

[6] Modelica—https://www.modelica.org/.

[7] FMI/FMU—https://fmi-standard.org/.

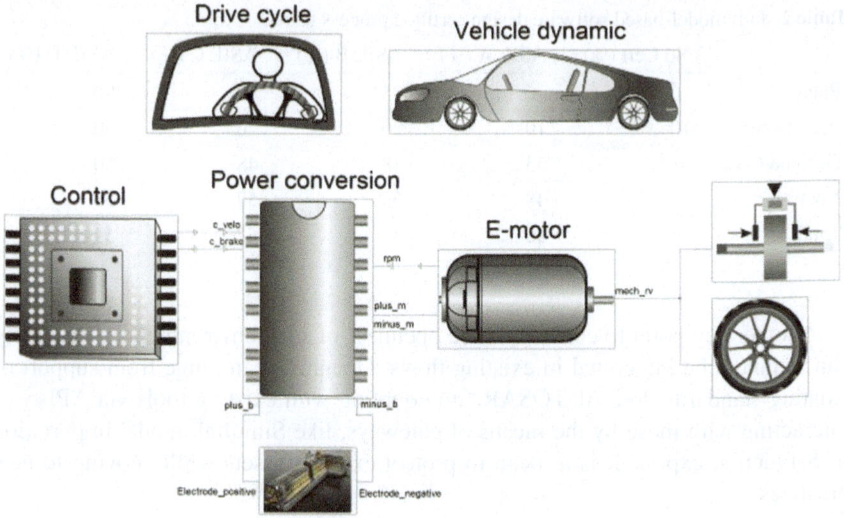

Fig. 8 BMS system simulation—Twin Builder schematic

Reduced-Order Models (ROMs), already mentioned Sect. 3. That technology permits the user to produce models that can simulate in fractions of seconds from 3D models that take hours (or days) to simulate, without loss of accuracy. ROMs can be applied to any kind of physics, and the latest development allows for generating ROMs for transient phenomena. They are used for instance to develop E-motors.[8]

Using the simulation, it is also possible to validate the nominal behavior of the system, but one can also ensure that the safety goals are fulfilled (see Fig. 9, showing results for WLTP class 3 profile,[9] for a 400 V battery) like that heat stays in the proper range, well as the voltage. System simulation also provides access to unexpected interactions as all components are brought together. Thanks to the various levels of granularity, one can have access to every detail of the components, which helps in understanding the observed behavior.

Such technology is so powerful that it now leads to the notion of "digital twin", where a complete virtual system like an electric engine or a full pump is coupled to the real assets for predictive maintenance.[10]

---

[8]https://www.ansys.com/-/media/ansys/corporate/resourcelibrary/article/engineering-e-motors-aa-v11-i3.pdf.

[9]https://en.wikipedia.org/wiki/Worldwide_harmonized_light_vehicles_test_procedure.

[10]https://blogs.sap.com/2018/06/05/sap-predictive-engineering-insights-enabled-by-ansys-digital-twin/.

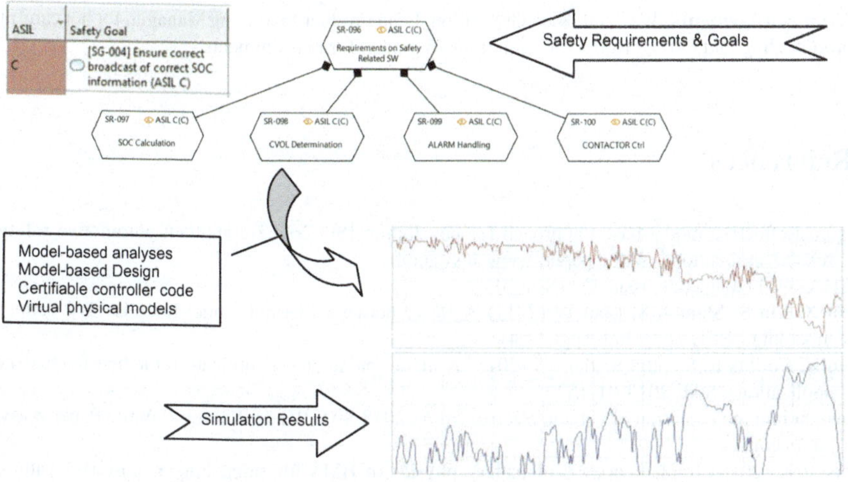

Fig. 9 Verification of safety goals with full system simulation

6 Summary

The "green" mega-trend is pushing for batteries that last longer and that can support long time discharge while giving high power to cover long distances. But battery development is a complex challenge as it combines complex multi-physics problems with an increasing need for safety due to the new power of the batteries and the forthcoming certification needs. Such certification already exists in industry areas like aerospace & defense or railways and transportation, but it will be mandatory at the highest level (ASIL-D) for autonomous vehicles in automotive industry. The reason is that a fully autonomous vehicle must be one hundred percent safe to be accepted by the market.

To achieve these goals, technologies already exist that have demonstrated their efficiency, being the 3D analyses of physics or certified model-based development. The good news is that all these technologies can be now integrated into a smooth flow, ranging from safety analyses down-to ISO 26262 certified code, with efficient and cost-effective process development. ROMs are the links between the software and in-depth analyses of the physical product and allow for a complete virtual simulation. Architecture exploration and design decisions can be made on trustable bases, and full final virtual system can be validated against functional and safety requirements. Such integration also helps the communication between teams that are part of the overall development. A great example has been demonstrated by Volkswagen who won the Pikes Peak Internal Hill Climb[11] with its electric vehicles.

[11] https://volkswagen-motorsport.com/index.php?id=2698&L=1

Acknowledgements We thank Kara Gremillion, Lead Product Marketing Manager, for her careful proofreading and Olivier Roll Lead, Solutions Engineer, for his comments.

References

A cost-effective model-based approach for developing ISO 26262 compliant automotive safety related applications. White paper, www.ansys.com

CHARGED magazine, issue Oct/Nov 2012

Hu X, Lin S, Stanton S, Lian W (2011) A foster network thermal model for HEV/EV battery modeling. IEEE Trans Ind Appl 47(4)

Hu X, Collins L, Stanton S, Jiang S (2013) A model parameter identification method for battery applications, SAE 2013-01-1529

Insight into the evolution of vehicle electrification: 2018 SAE Reader Survey—White Paper www.ansys.com

Lee T-K (2018) Implementation of battery physics in BMS for safer, longer, powerful battery operation. In: 5th international conference automotive battery management systems, 18–20 Sept 2018

Trends in Power Electronics Impacting E-Mobility/SiC as Key Enabler for Greener Driving

Manuel Gärtner

Abstract Silicon Carbide (SiC) components have been used for a number of years in industrial applications and now are progressively entering the automotive segment. The first devices of this wide-bandgap material to be used in vehicle applications were SiC diodes in on-board charger (OBC) applications. The excellent SiC diode switching performance is ideal for the freewheeling path in the power factor correction stages. The market for automotive grade SiC components is showing impressive growth aligned to the rise in electric vehicles with SiC MOSFETs being used in traction inverters as well as OBC or DC-DC converter applications. This paper explains the significant advantages that wide-bandgap technologies have compared to well established Si technologies. This applies especially for the total cost of ownership approach, not considering the single device only but the complete application or interaction between different components. When a new semiconductor material is introduced into the automotive sector, the stringent qualification requirements mean that a special focus is needed on the manufacturing strategy. This paper also describes a manufacturing strategy designed to serve the automotive market, the key reliability factors and how improvements have been achieved.

Keywords Power electronics · Electric vehicle · Silicon carbide · SiC · Efficiency · Automotive · Traction inverter · On-board charger · Wide-bandgap

1 Introduction—Electro Mobility Is Gaining Momentum

Electric vehicles are evolving from a niche solution to a reasonable alternative for mobility in the 21st century. There are many factors that are promoting vehicle electrification, from improved technologies, increasing charging infrastructure and environmental legislation.

M. Gärtner (✉)
STMicroelectronics Application GmbH, Bahnhofstrasse 18, 85609 Aschheim-Dornach, Germany
e-mail: manuel.gaertner@st.com

J. Langheim (ed.), *Electronics Components and Systems for Automotive Applications*, Lecture Notes in Mobility,
https://doi.org/10.1007/978-3-030-14156-1_9

1.1 Legislation

The Paris Climate Agreement, signed in 2016 by 195 countries, aims to create a lower carbon future. Part of the solution is reducing vehicle emissions. The Paris Declaration on Electro-Mobility targets over 100 Million electric vehicles by 2030. Reinforcing these efforts, CO_2 emission legislation in Europe targets 95 g of CO_2/km for passenger cars with carmakers facing fines if they fail to achieve these targets.

Diesel power had been championed and subsidized by many states as having lower carbon emissions than petrol but studies[1] into the health risks of diesel air pollution, due notably to nitrogen oxide and particle emissions, have disputed that benefit. Now cities are considering bans of diesel-powered vehicles completely and several countries including Norway (2025), India (2030), France and UK (2040) are phasing out sales of petrol (gasoline) and diesel-powered vehicles completely.

1.2 Incentives

Government incentives to support the lower carbon future are widespread. Innovative schemes exist across the world based upon tax credits and VAT exemption.

1.3 Practical Alternatives

Technological advances are also making electric vehicles a practical alternative to Internal Combustion Engine (ICE)-based vehicles. Cheaper battery technology and improvements in range and performance combined with a growing charging infrastructure are influencing consumers and convincing them that a greener option can be both an economic and ecological stance.

- Battery costs have fallen by a factor of four since 2009, with energy density increased six fold[2]
- Global Charge Station progression (EVSE[3]), has grown 450% year-on-year in 2016.

[1]https://www.rcplondon.ac.uk/projects/outputs/every-breath-we-take-lifelong-impact-air-pollution.

[2]https://www.iea.org/publications/freepublications/publication/GlobalEVOutlook2017.pdf.

[3]Electric vehicle supply equipment.

1.4 Many Forms

Vehicle electrification comes in different forms, from the battery electric vehicles (BEV), through hybrids and mild hybrids. It also includes traditional combustion engine vehicles with start-stop functions at the 12 V level and vehicles equipped with extended start-stop functions and certain electric drive characteristics up to mild-hybrid vehicles with 48 V networks. Vehicles equipped with such systems offer drivers their first electric vehicle experience, and already have a positive impact on the environment.

Apart from an acceptable range, users of electric vehicles will also expect the same level of reliability in EVs as in traditional cars, at the same cost. This will require important investment over the next decade in order to reach the same maturity levels for EV components as has been achieved for ICE components. There are many research predictions into the rise of electric vehicles, and all agree that the next decade will see a strong increase. Some market researchers sees EV sales rising up to 35% of all vehicles by 2040, with significant growth starting around 2025.

To make this happen a number of technological, technical, industrial and logistical issues have to be solved.

This paper will focus on the technological solutions, looking at electric vehicles with high voltages. The latest developments in power electronics for electric vehicles will be discussed, with a particular focus on the benefits of Silicon Carbide technology.

2 Wide Bandgap Material

Wide bandgap material is the foundation of power electronics of the future. The below figure of merit (Fig. 1) shows the principle advantages of wide bandgap materials Silicon Carbide (SiC) and Gallium Nitride (GaN) compared to Silicon (Si) technologies.

These wide bandgap technologies are almost everywhere better than Si. Operating temperature are higher, switching losses as well as conduction losses are lower and heat dissipation is easier to manage.

Unfortunately, these materials are more difficult to process and manufacture compared to than Si, which has delayed their introduction until recently.

SiC is a very hard material, crystal growth needs a lot of time and energy and defects have been numerous in the past heavily reducing yields.

GaN is in fact a layer of GaN on a Si wafer with different temperature coefficients, which makes the heating and cooling quite sensitive.

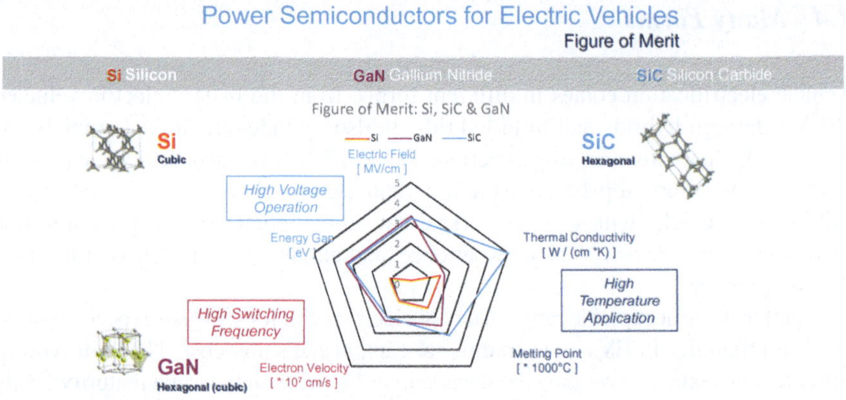

Fig. 1 Figure of merit of Si, SiC and GaN

3 Industrial Deployment of New Technologies

Introduction of new technologies into a market demanding high quality, long and reproducible lifetime at competitive costs is a challenge. Today, the difficulties in using wide bandgap materials in industrial production have been overcome for SiC and are advancing considerably for GaN at present.

The supply chain of SiC is becoming more robust, the cost of the raw material is decreasing due to competition and the quality of the material and the process is improving steadily. These factors are allowing the use of SiC in automotive applications with high quality and competitive cost requirements. Additionally, there will be a strong increase in volume driven by increasing power needs of automotive applications, which will drive down costs by the economy of scale.

3.1 Industrial Deployment of SiC

At STMicroelectronics, we have been producing SiC products for a number of years. It started first with 2″ wafers, moved then up to 6″ for the future it is expected to be available on 8″ wafers (Fig. 2) (Gärtner et al. 2016).

The epitaxial layer, which is critical for SiC devices, is done in-house in STMicroelectronics. This is a mandatory in order to assure the highest quality, highest yields and the ability control higher currents.

Our experience in terms of reliability, quality and in particular automotive quality is reaching similar levels as to those of Si devices. Deployment of SiC technology in electric vehicle (EV) markets is increasing strongly.

This will certainly have an impact on the cost of ownership of cars from 2021 onwards with an immediate effect on the image of electric vehicles. In terms of total

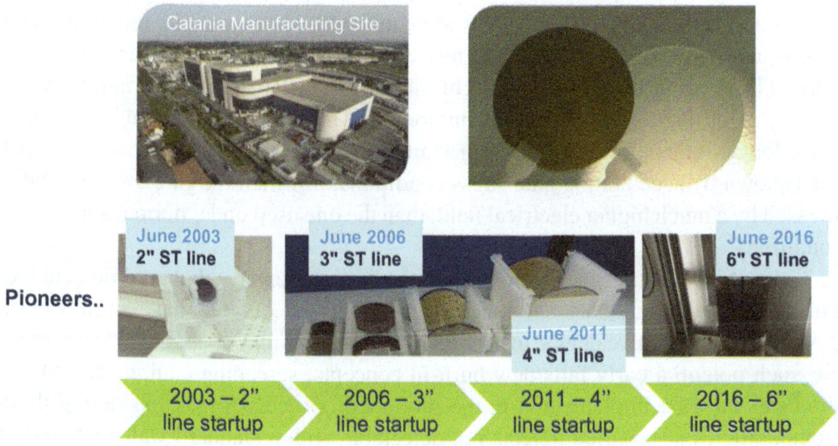

Fig. 2 Evolution of wafer sizes for SiC production at ST over time

cost of ownership, SiC will be an important factor in making EVs competitive with conventionally powered vehicles.

After gaining experience serving the less stringent industrial segments, STMicroelectronics has invested heavily in technology and production facilities to serve high volume car platforms. Consequently, the Catania SiC production line capability has been continuously extended and the wafer starts per week have been significantly increased (See Fig. 3).

Technical challenges like reliability of the final device and defect density of the SiC substrates are two of the key elements that required attention to allow the new material to break into the automotive industry. This industry segment is characterized by very demanding quality targets, to avoid problems on the production line or on the road. Knowledge built up in defining and improving previous SiC MOSFET generations has been used to create a more robust technology, fit for automotive applications.

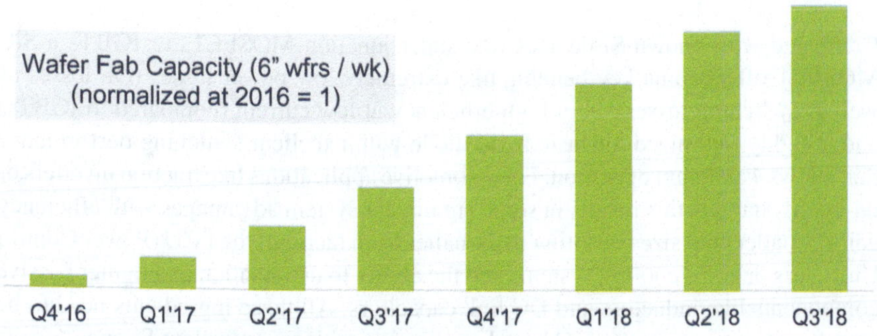

Fig. 3 Evolution of Catania SiC wafer fab capacity

Gate oxide integrity has been improved significantly. For example, reliability tests on the latest SiC device generation, has shown minimal threshold parameter drift after HTGB (high temperature gate biasing) or HTRB (high temperature reverse biasing). Optimizing the process technology and process equipment also leads to a much longer time before encountering a failed device (mean time to failure). This has been shown by accelerating gate stress conditions, in which the gate oxide has been stressed by a much higher electrical field, than the one used under normal application conditions.

Today, SiC substrates still show a relatively high density of crystal defects like carrots, triangles, scratches and macro steps. Such defects may have an impact on the lifetime of an active device like a SiC MOSFET and may cause an instant failure. For such potential early fails new burn-in concepts, screening equipment and test strategies have be implemented in the production process flow. Considering these methodologies also buried defects not detectable with a standard wafer fab end of line test (EWS/electrical wafer sort) can be screened out.

3.2 *System Aspects and Packaging*

Packaging for wide bandgap materials is in its infancy and offers great potential for improvements. Current packaging solutions are mainly based on traditional Si experience. Higher temperatures will need further R&D and due to the much faster switching capabilities of wide bandgap materials, parasitic parameters like the leakage inductance in the DC link, will become key areas of research in new packaging technologies. This is also true for the passive components (e.g. the DC link cap) in a system. Solutions exist, but cost and size are still important. New architectures might offer the opportunity for further optimization.

4 SiC Value Proposition and System Advantages

Compared with known Si devices like super junction MOSFETs or IGBTs a SiC MOSFET offers some key benefits like extremely low power losses (On losses as well as switching losses), very high efficiency at low current (compared to IGBTs) and a highly optimized intrinsic body diode with excellent switching performance that allows 4 quadrant operation. For automotive applications like traction inverters or on-board chargers this results in some significant system advantages with efficiency gains, smaller chip sizes resulting in a smaller from factor of the PCU (Power Control Unit), less onerous cooling systems and the ability to use smaller, and lighter passive components like inductors and DC link capacitors. All these ingredients need to be considered to compare the SiC based application with an equivalent Si one.

The above-mentioned system advantages (Fig. 4) must be considered in a total cost of ownership approach, as today, the cost of producing SiC MOSFETs is higher than

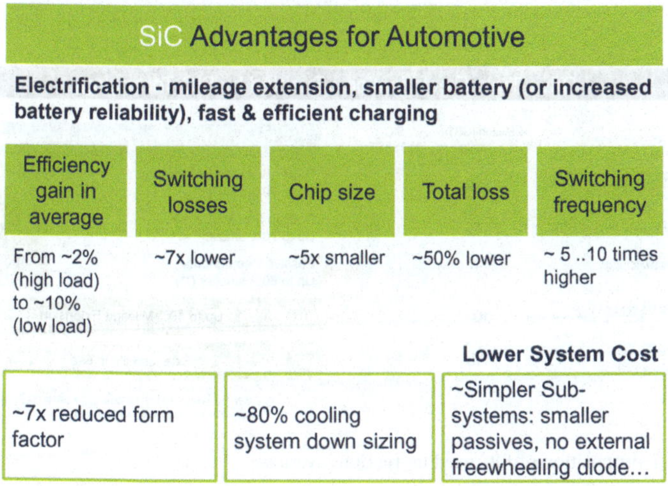

Fig. 4 Advantages of SiC components used in automotive applications

IGBTs. Let us consider an example application that could be built using SiC or IGBT technology, the traction inverter. One of the automakers design goals is to improve the e-powertrain efficiency for typical driving cycles. IGBTs are well suited for high load conditions but at partial load, the predominant operation point during normal driving of a car, SiC MOSFETs have a significant advantage. A second advantage comes from the use of power switches in parallel. As an IGBT is characterized by the knee voltage which does not decrease when used in parallel. A SiC MOSFET however, behaves like a resistance and the usage of devices in parallel results in lower on-resistance. Comparing efficiency simulations, to benchmark a Si based IGBT plus freewheeling diode module versus a full SiC MOSFET shows an efficiency gain of the SiC solution of around 3–5% for a 400 V DC bus or up to 8–12% for an 800 V DC bus. The greater efficiency means improved mileage or the car manufacturer may decide to maintain the range, but can use a smaller battery, and decrease the cost of the vehicle. Considering that battery cost is around 30–40% of the production cost of a car, it is evident that investing in high efficiency semiconductors is cost effective. Smart and efficient handling of the invaluable charge stored in the battery will help offset overall vehicle costs. Semiconductor power losses inside the traction inverter generates heat, which requires cooling systems that also add weight and hence reduce mileage. Smaller and lighter passive components also have an impact on cost and efficiency. SiCs higher operating temperature and lower losses combine to reduce the end users cost per kilometre. Efficiency gains can be in the region of 1–2 kWh saved per 100 km.

SiC based on-board charger applications, offers the possibility to run at much higher pulse width modulation (PWM) frequency in the power factor correction (PFC) stage as well as in the (half) bridge LLC stage, when compared with Si based solutions. In this application (Fig. 5), the motivation is not primarily the efficiency

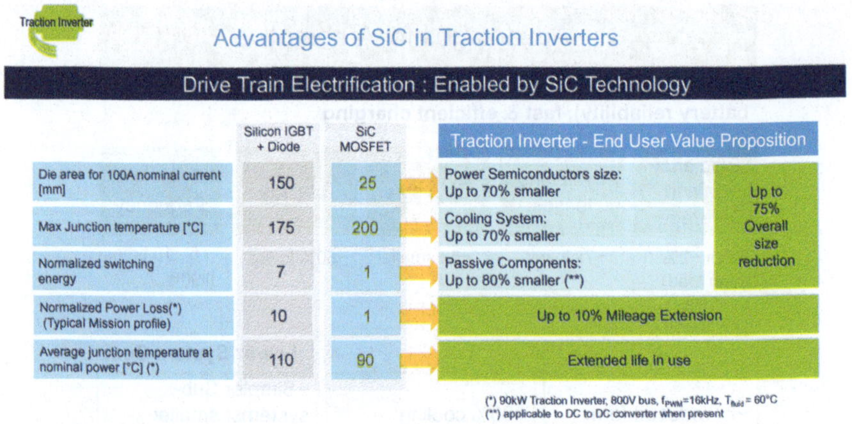

Fig. 5 Value proposition of SiC used in traction inverters

gain, but the significant impact on the form factor of the power control unit (PCU). In a second case study (Beranger 2015), that was performed on a 3 phase 11 kW on-board charger (Fig. 6), a significant volume reduction of the PCU was achieved. In addition, the size of bulky passive components like coils and capacitors is also strongly decreased, which also decreases the costs for these components. The case study shows a volume reduction from about 4.5 l for the Si based solution to a value of about 2 l, which can be considered a key advantage for the wide bandgap based solution.

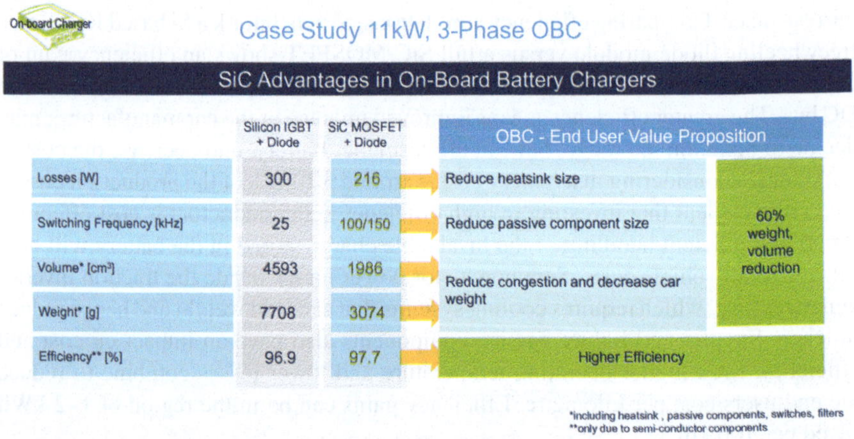

Fig. 6 Value proposition of SiC used in on-board chargers

5 SiC Innovation at STMicroelectronics

The automotive industries adoption of SiC has been much faster than expected. In 2017 more than 2 Million SiC MOSFET units have been shipped to automotive customers and the forecast for the year 2018 is already 10 times higher. SiC diodes have been used for several years. To serve the different DC bus voltages, MOSFETs and diode product families are available in 650 V technology for 400 V DC bus based systems and a 1200 V for the upcoming 800 V boardnet. To address the needs of the customer, the devices in each voltage class are scalable in performance. For MOSFETs, a continuous shrink in Rdson has been achieved at each technology node evolution (Fig. 7). This is a key parameter especially for traction inverter applications. From the 1st generation Rdson has been divided by a factor of 2 for the 2nd generation. The 3rd generation with production in 2019 will halve this key parameter again.

For packaging different options are available or are under development. The components are available as bare die to be used in customer specific packaging technologies, in standard trough hole (TO247) or SMD (H2PAK) packages or as modules (Fig. 8). The technology as well as packages is available as "ready to use" standard components or can be customized to customer requirements. With EV architectures evolving all the time, this flexible approach allows customers to rapidly develop solutions built on proven technology.

6 Conclusion

The automotive market is undergoing a disruptive change due to strongly growing demand for electric vehicles, both battery powered and hybrid solutions. New wide-bandgap devices are offering major advantages, especially when considered in terms of a total cost of ownership approach. We see very strong growth for SiC MOSFETs in traction inverter, on-board charger and DC-DC converter applications. The motivation to introduce this new material is the optimized efficiency for the traction inverter, which can lead to an extended mileage of 10% or to a downsizing of the battery capacity of the same order.

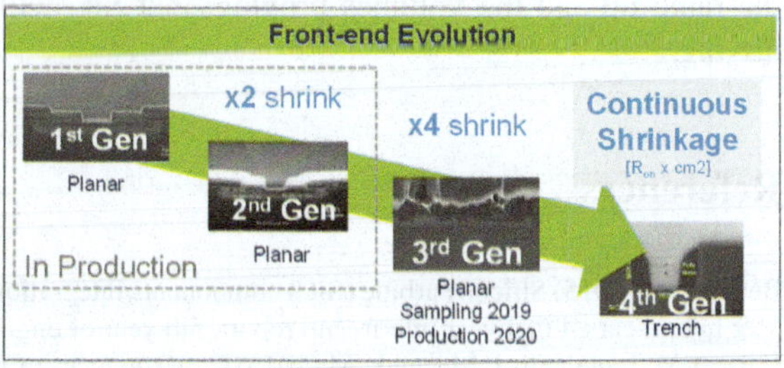

Fig. 7 SiC MOSFET technology roadmap

Fig. 8 Different packaging solutions

In applications running at high frequencies, the benefits include the downsizing of bulky passive components and the form factor of the power control units. Introducing SiC components to the automotive world is a difficult task, with challenging requirements for manufacturing and quality. STMicroelectronics has successfully introduced automotive grade SiC components in high volume applications, with increasing numbers of devices "on-the-road" in electric vehicles. The speed of SiC technology introduction in the automotive market is much faster than expected; with the ramp-up and the volumes proving, that SiC will be a key enabler of greener driving.

References

Beranger Q (2015) Silicon carbide (SiC) components integration in an on-board charger for electric & hybrid cars, Final year intern-ship report. 5th year of engineering school at Polytech Tours

Gärtner M, Langheim J, Macauda M (2016) Consequences of new power electronics technologies on the market evolution of EVs, e-Motive, Schweinfurt

Powering Up Electronics—Latest Developments and Concepts for Packaging of Electronics in Automotive Systems

Johannes Stahr and Mike Morianz

Abstract During the funding project EmPower a new embedding concept was developed for integration of power components. The embedding technology shows promising potential for automotive applications regarding high miniaturization potential as well as good electrical and thermal performance. Additionally the concept provides good reflow stability and thermal shock behavior. This could be realized with the demonstrators of the EmPower project and single chip test vehicle for deeper investigation of the potentials of this embedding technology according AEC-Q101 requirements. High robustness on power cycling was an outstanding exploration result.

Keywords Mild HEV · Full hybride

1 Introduction

To improve the efficiency in modern cars on the way to electromobility the power density has to be increased without limitations on performance and reliability. The implementation of power semiconductors like MOSFETS, IGBTs or diodes directly into the PCB is a very promising approach to fulfill these needs. AT&S successfully used the expertise with its ECP (Embedded Components Packaging) technology for the implementation of efficient power packages and modules. This made it possible to reduce the space required for power packages by up to 50% with correspondingly higher power density. In addition, it showed that very good results are achieved in terms of switching behavior, heat removal and power cycling robustness.

J. Stahr (✉)
R&D, AT&S AG, Fabriksgasse 13, 8700 Leoben, Austria
e-mail: H.Stahr@ats.net

M. Morianz
R&D, AT&S AG, Fabriksgasse 13, 8700 Leoben, Austria

J. Langheim (ed.), *Electronics Components and Systems for Automotive Applications*, Lecture Notes in Mobility,
https://doi.org/10.1007/978-3-030-14156-1_10

2 Electromobility Needs Innovative Power Concepts

Besides autonomous driving and connected cars the automotive industry is driven by the megatrend electromobility, forced by government regulations in the main regions. The motivation to introduce electromobility is based on the needs to reduce the emissions and the fuel consumption of the fleet. This can be done in several steps. The first step is to improve the efficiency of the conventional ICE (Internal Combustion Engine) with technologies like start-stop and modern EPS (Electric Power Steering) systems. In addition, the energy efficiency of subsystems like electric pumps, electric fans and the power distribution have to be improved. The most important and significant step is the drivetrain electrification with subsystems like inverter, onboard charger and battery management.

The EU defines a reduction of the CO_2 emission down to 95 g CO_2/km on average for the fleet until 2021. At the moment a further reduction to about 60 g CO_2/km until 2030 is discussed. China—the leading region bringing electromobility on the street—for example plans to reduce the average fuel consumption down to 5 l/100 km until 2020. All these goals can only be reached with high-voltage electrification, based on high-efficient power semiconductors and power packaging (Fig. 1).

On the way to complete electrification there are different stages with increasing efficiency. Mild HEVs (48 V) uses improved recuperation and boost features with start-stop functions and energy recovery. This can lead to about 15% energy efficiency improvement compared to a typical ICE. Full-Hybrid (FHEV) added features like coasting, sailing (driving with turned-off motor) or boosting which leads to about 30% efficiency improvement. With the Plug-in Hybrid (PHEV) the high-voltage

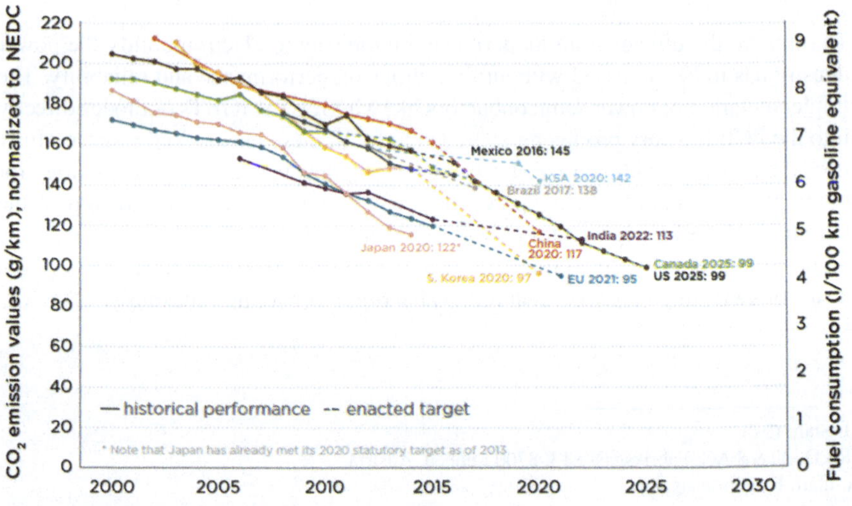

Fig. 1 Historical fleet CO_2 emissions and current standards. *Source* iCCT

electrification starts. This approach enables a CO_2 reduction of up to 80%. The final aim is a completely battery supplied car (BEV) with no CO_2 emission .

To realize high-efficient electromobility drive drains both high performance power semiconductors and efficient and reliable PCBs with innovative power packaging technologies are needed .

3 The EmPower Project

The Eureka Catrene "EmPower" project with the goal to create embedded power components for electrical vehicle applications was running from May 2013 to February 2017. The project coordinated by AT&S included ST Microelectronics, Atotech, ILFA, Continental, TU Wien and TU Berlin as further members. The project covers the complete Embedded Power tool chain from wafer supply, wafer plating, wafer backend, die embedding, power modules up to system development and system simulation.

Interconnections between the power devices and wire bonds on Direct Copper Bonded (DCB) substrates are a primary source of parasitic inductances in today's power packages, leading to significant switching losses, limited lifetime and reliability issues due to the high-power loss density. In contrast, the EmPower packaging concept contains copper interconnections with large cross sections between the power devices and the package pads by short copper-filled via and large copper cross section. This technology allows to place the necessary copper connections and pads on both sides of the package in an arbitrary manner, thus satisfying the needs for both, high electrical and high thermal conductivities.

The aim of the EmPower project was to develop a packaging concept for power electronic current regulators suitable for the operating conditions and reliability requirements encountered in the field of electromobility. The primary fields of application are power converters for drive motors, charging devices and power electronic controls in auxiliary units such as power steering, brake system, etc. (Fig. 2).

The innovative advances of the EmPower concept consist on the one hand in a shortening of the heat removal paths, via which the switching losses of the power semiconductors (IGBT, MOSFET, diode) are removed to the heat sink and whereby the thermal load of the assembly is significantly reduced under a load. On the other hand, the concept sees the power semiconductors being coated for the first time on both sides with copper and embedded in a printed circuit board construction using a new low-temperature silver sintering process.

Lastly, making the shortest possible electrical connections between the power semiconductors requires minimizing the parasitic line impedances and significantly improving the switching performance.

In the course of the project, three different demonstrators for the power ranges 50 W, 500 W and 50 kW were developed.

The embedded components (PARSEC, Planer Surface Embedded Components) are positioned with backside directly into the Cu-layer of built-up. This leads to an

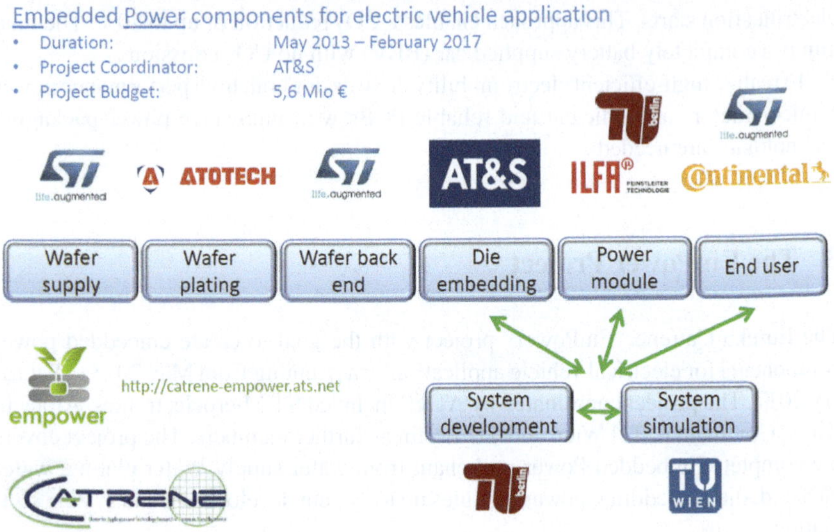

Fig. 2 EmPower was Catrene funded EU project

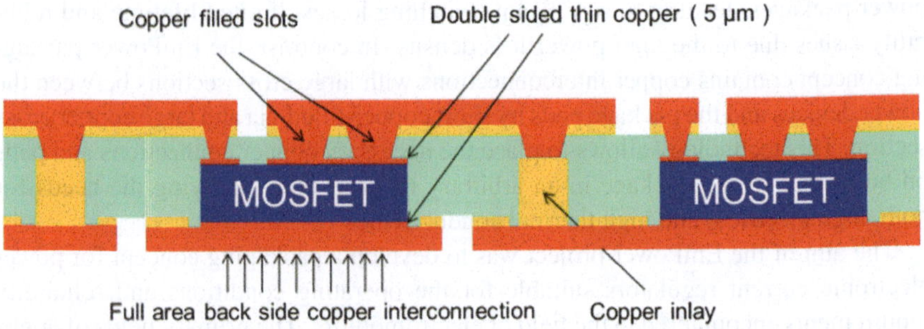

Fig. 3 Schematic of the embedded power concept

improved switching performance, reduced inductance and improved thermal resistance (Fig. 3).

As part of the EmPower programme, among other things, the equipment and processes were provided for realizing double-sided copper-coated power components. This means that the complete process chain for embedded power components and new power packaging solutions is now available—namely, from wafer fabrication, to the embedding of chips, through to the assembly of power components and modules.

4 The Process

At the heart of the EmPower implementation are so-called power cores as function blocks for power modules. The innovation in terms of the new power packaging concept entails the embedding of power devices such as IGBTs, MOSFETs and diodes as chips with thicknesses from about 300 μm down to as low as 100 μm in glass fibre reinforced epoxy layers with a total thickness of 450 μm down to only 150 μm.

The aim here was to create relatively large connection areas, on the one hand to achieve low impedance and on the other hand optimized heat removal. The thermal behavior is also optimized by embedded copper inlays and their contacting with galvanic connections on the upper and lower side of the power component. The power module integrates IMS (Isolated Metal Substrate) layers on the upper and lower side with the power core (with embedded MOSFETs and copper inlays) in between. Cost-effective and effective thermal management is made possible by embedding the power semiconductors in a module that can remove the heat on the upper and lower side.

This technology allows power semiconductors to be embedded in such a way that the rear of the power module provides a complete copper surface for the connection and allows chips of up to 10 mm × 10 mm to be embedded. The contacting of the pads on the upper side of the components is done by means of micro-vias or full area interconnection.

5 Demonstrators

In order to evaluate the scalability and performance of the new power packaging technology, several demonstrators in the range from 50 W to 50 kW were developed. For the 50 W power range, the corresponding demonstrator takes the form of a power package (50 W, 100 V, 20 A) with an integrated high-speed rectifier diode. This achieves a miniaturization of more than 50% compared to a comparable standard performance package. In particular, the thickness is much smaller than with a standard D2PAK package.

For the 500 W power range (12–48 V, 20 A), a power module was selected for a pedelec application (brushless DC motor drive). This demonstrator shows the concept of a power module with double-sided cooling and an integrated B6 bridge as a DC/AC converter (Fig. 4).

For the 50 kW power range (250–450 V, 200 A) a power module for hybrid/electric vehicle applications was developed.

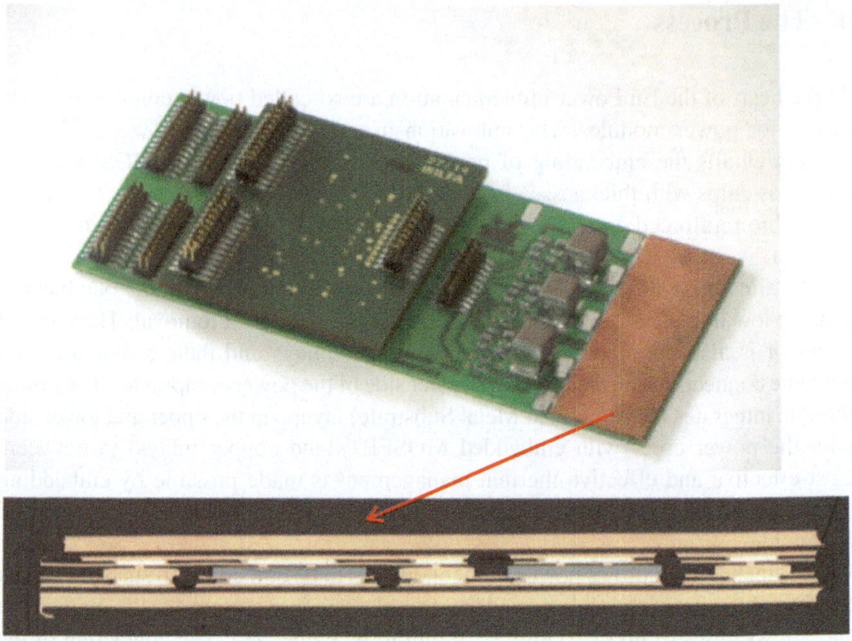

Fig. 4 500 W demonstrator—size reduction in x, y of about 50% compared to a SMT solution

6 500 W Demonstrator

The 500 W demonstrator for pedelec applications uses the power module developed by EmPower. AT&S' contribution focuses on the power core with embedding of the MOSFETs and the copper inlays. In this way it was possible to reduce the electrical and thermal resistance as well as the inductance. The reduction of the overvoltage indicates reduced switching losses and finally possible faster switching.

The 500 W demonstrator with B6 bridge configuration uses double-sided cooling. For the purpose of benchmarking, the 500 W demonstrator was compared with a standard power module in SMT technology.

When investigating the switching behavior, a half bridge was considered, and double pulse measurements were performed for different gate resistances. This showed a significantly improved switching behavior for the embedded power solution. This is because the embedding concept allows the individual components to be placed with very small distances, which reduces the inductance in the switching cells. This allows the overvoltage to be lowered when switching off the MOSFET, which in turn permits faster switching or lower switching losses (Fig. 5).

The thermal behavior of the 500 W demonstrator was also investigated. This was done by a comparison of junction-to-heat sink thermal resistances of different cooling cases with those of a benchmark module fabricated in SMT. A corresponding test setup confirmed the simulated results and shows that the module satisfies the design

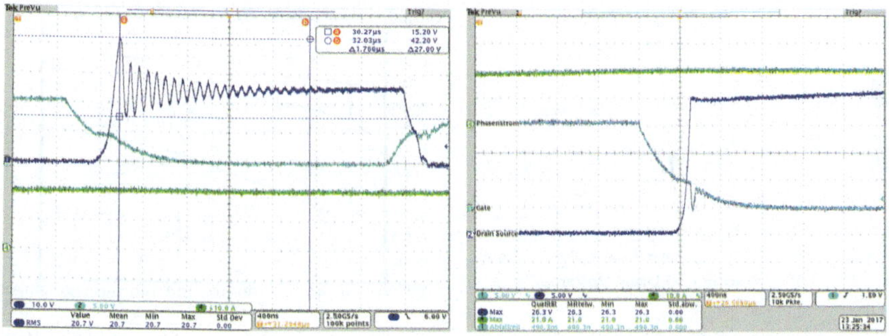

Fig. 5 Switching performance of the 500 W demonstrator (right) in comparison to SMT technology (left). Parameters: Switch off behavior with a gate resistance of 51 Ω (Green: Phase current (20A), Light blue: Gate voltage (5 V), Blue: Drain. *Source* Voltage (24 V)

specifications. In addition, temperature shock tests showed that even at 1000 TCT (Temperature Cycling Test Cycles at −40 °C/150 °C), the thermal resistance of the MOSFETs did not increase significantly and there were no cracks or delaminations in the area of the power core, IMS, or connections. MSL3 (moisture sensitivity) and reflow tests also proved the high robustness of the demonstrator.

7 Electrical and Thermal Performance

A test board with embedded 40 V N-channel MOSFETs, supplied in tape&reel was used to evaluate the reliability and electrical performance of the Embedded Power technology. The MOSFET size in this case was 6.34 mm × 4.6 mm × 0.210 mm with a 6 μm Cu pad termination on top and bottom side. The gate pad and source pad on top side were connected via Microvia. The drain pad on bottom side was connected with full area copper-plating. The SEM/EDX analysis of the MOSFET and the Microvia gate contact showed high reliability.

To examine the electrical performance of the embedded MOSFETs a power device analyzer (Keysight) was used. All measured key specifications from breakdown voltage to the body diode forward characteristic were inside of the specification. For the efficiency the $R_{D(on)}$ is a key parameter. Compared to a similar SMT package the values are 0.4 mΩ lower (Fig. 6).

A HTRB test according AEC-Q101 (reverse bias) was used to examine the reliability against humidity and temperature and passed 1000 h. In addition, the R_{DSon} was examined for HTRB (High Temperature Reverse Bias) and HTGB (High Temperature Gate Bias). Both showed only very small changes of the on-resistance. The same for the TCT test according AEC-Q101 (Fig. 7).

Active power cycling is an important test for power electronic components and modules. Tests with the 500 W demonstrator (dT = 100 K, 3 s interval for switch

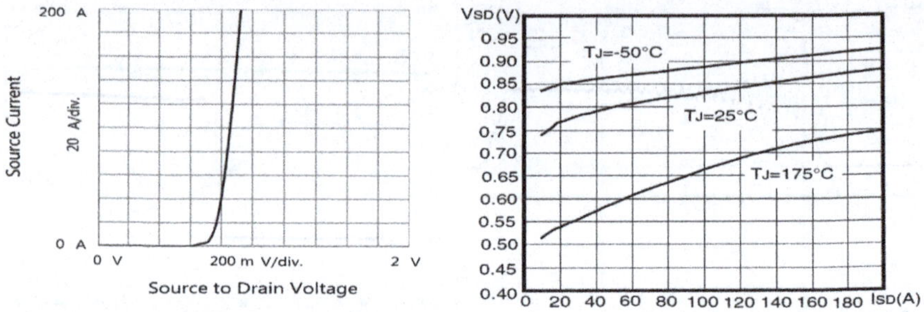

Fig. 6 Measured values of all key specifications like the body diode forward characteristics (left) fits exactly to the data sheet (right)

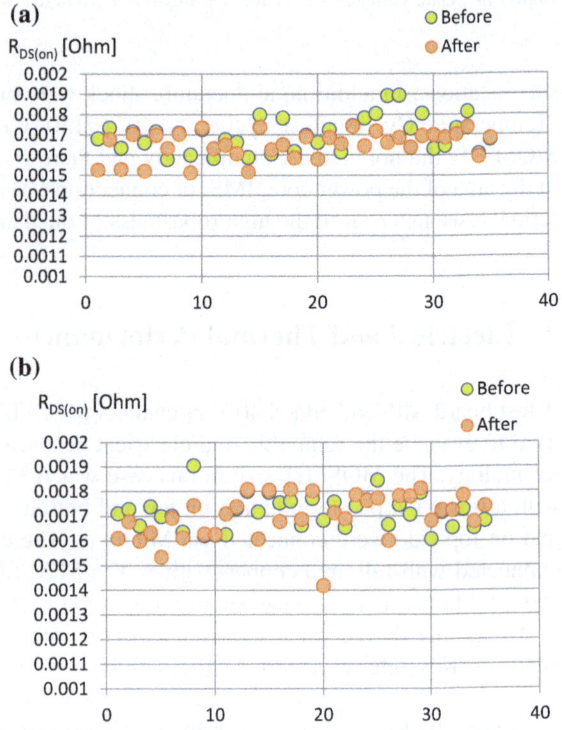

Fig. 7 **a** R_{DSon} before and after HTRB (AEC-Q101), **b** R_{DSon} before and after HTGB (AEC-Q101)

on and switch off) passed more than 300,000 cycles. Other projects with various temperature differences and stress types shows the successful pass of 600,000 cycles and more.

8 Conclusion

The Embedded Power concept have proved to enable miniaturization, improved switching performance and improved thermal resistance. Reliability according AEC-Q101 have shown in different tests. Power cycling results of different projects show high robustness of the embedding technology.

Since 2010, active and passive components have been integrated into circuit boards for low-power designs and have already been mass-produced. The AT&S ECP® technology for volume production meets the high-quality requirements. Against this background, there is no limitation to integrate power components directly into the circuit boards in order to take advantage of the benefits of miniaturization.

Reference

http://catrene-empower.ats.net/publications/

Part IV
Autonomous Driving, IoT and Data Processing

"How Good Is Good Enough?" In Autonomous Driving

Hans-Peter Schöner

Abstract A new approach for quantifying "good enough" for the behaviour of autonomous vehicles (AVs) is presented here. It is deducted from general behaviour guidelines for (human) behaviour in traffic, and especially the wording of §1 of the German Traffic Code ("Straßenverkehrsordnung, StVO") is taken into account. The German traffic code generally requires constant *caution* and *mutual consideration*; in a second paragraph, it more specifically requires, that "no one is **harmed**, **endangered** or unnecessarily **hindered** or **bothered**." The here presented approach proposes a way how to specify and finally quantify those key words. These quantifications can be derived from normal (i.e. safe) traffic practice, which has ever since been established based on the human capabilities to cope with traffic situations. First, this quantification defines the guidelines for a safe own behaviour of an AV, for normal (avoid hindering or bothering others, but never endanger them) and extraordinary (never harming others) traffic situations. It establishes behaviour rules and thus testable performance guidelines for a **single vehicle**. Second, based on the fact, that *everybody* should comply with §1, certain behaviour of traffic partners can be trustfully *expected* in any **interactive traffic situation** (even an AV should not be harmed or endangered by any other traffic participant), but a safe and robust reaction on (rare) expectable situations (collision free, if only being hindered or bothered by other participants) needs to be ensured. The quantifications can help to define the functional design space for AV behaviour in normal traffic scenarios, with the goal of "not endangering" traffic partners. This leads to more robust behaviour than the "no collision" goal for extraordinary situations. The concept allows to deduct testable performance goals in reference situations, with reasonable passing criteria. The quantification of the key words is mainly based on the **reaction times** for safety-relevant actions in traffic scenarios, in combination with **manageable reaction patterns**. Human reactions are generally limited by necessary perception, interpretation, reasoning and action times. Thus, the requirement for AVs must ensure equal or better total performance. This cannot (and is not required to) be tested in every conceivable

H.-P. Schöner (✉)
"Insight from Outside"—Consulting, Ringstr. 2, 72119 Ammerbuch, Germany
e-mail: hans-peter.schoener@gmx.net
URL: http://www.ifo-consulting.com/

J. Langheim (ed.), *Electronics Components and Systems for Automotive Applications*, Lecture Notes in Mobility,
https://doi.org/10.1007/978-3-030-14156-1_11

situation; but typical traffic reference situations must be agreed on, which define borderline cases. Some outer conditions must be summarized by testing for the extreme case. Any testing procedure should include verification of the ability for autonomous stopping; for the worst case of total sensor loss under extreme weather conditions, the "**Blind stopping procedure**" is proposed. It must be accepted that there exist conceivable situations, in which a single traffic participant will not be able to avoid an accident (loss of own controllability), once a situation has evolved to a certain criticality. The quantifications help to draw a limit line between "avoidable" and "unavoidable" (force majeure, natural disaster) accidents. In order to stay away as much as possible from getting into such uncontrollable situations, an AV needs to demonstrate a perception of "**level of danger**" and a sense for its own capabilities ("**self-awareness**"). Testing for these skills should be part of release and certification procedures.

Keywords Autonomous vehicle · German traffic code · Certification · Verification · Validation · Endangering · Collision avoidance · Reaction time · Reaction pattern · Blind stopping procedure · Self-awareness · Level of danger · AV driver's license

1 Background and Motivation

Since several years an international discussion on the testing, verification and certification of autonomous vehicles (AVs) is going on (Schöner 2014; Steininger et al. 2015). In Germany, the PEGASUS-project[1] is on the way with the goal to define "how good is good enough?" and "how can we verify this?" with respect to the introduction of AVs. Internationally, many other activities and conferences are dealing with this topic.

One of the key reasons for the long-lasting discussion is the insight, that the often-announced goal of an accident-*free* road traffic with autonomous vehicles is theoretically impossible to reach: since there are many independently acting traffic participants, and since the traffic is exposed to uncontrollable environmental conditions, it cannot be designed to be 100% accident-free.[2] Assumptions for the operational design space of AVs need to be made, in order to come to a feasible design solution, with much fewer accidents to be expected than in the actual purely human-controlled road traffic. These assumptions imply, however, that there might be accidents, which will occur predictably under certain (rare, but conceivable) conditions. Product liability legislation requires, that nobody should provide a product which is undue unsafe by its nature; in order to resolve this apparent contradiction, a generally accepted guideline for designing a *sufficiently safe* AV behaviour in traffic is indispensable.

[1] https://www.pegasusprojekt.de/en/about-PEGASUS.

[2] https://www.mobileye.com/responsibility-sensitive-safety/rss_on_nhtsa.pdf.

One important aspect is the interaction of traffic participants. A safe traffic can only be designed if all participants comply to certain rules. Recently,[3,4] a complete set of formalized design rules, which mimic human behaviour in traffic, are published; the design rules are called "Responsibility-Sensitive Safety (RSS) model" by the authors. They explicitly state, that "the RSS model contains parameters whose values need to be determined through discussion with regulatory bodies and it would serve everyone if this discussion happens early in the process of developing autonomous vehicles solutions".

The approach discussed in this paper has been independently developed from the RSS model, but the ideas are close to each other and complement each other. This paper strives to provide a new view on how to define the parameters for defining "good enough", being derived from normal traffic practice and specifically based on §1 of the German Traffic Code ("Straßenverkehrsordnung, StVO"), which is the general guidance for (human) behaviour in traffic. Besides this, this paper gives suggestions on how to test AV skills and monitor AV behaviour in order to provide evidence of safe behaviour for release and certification purposes.

2 The Goal of Defining "Good Enough" in This Context

This paper focusses on the creation of guidelines for fair dealing with rare challenging situations caused by the environment of AVs. These situations present a continuous spectrum from easy manageable to impossible to control, even for human drivers, but also for autonomous vehicles.

They can be divided specifically into two categories:

- suddenly emerging, challenging environmental situations ("emergencies")
- extreme and/or unexpected behaviour of other traffic participants (behaviour outside of "law and order")

These challenges do not derive from specific technical solutions, but they are caused by the properties of the application "public road traffic". They deal with the situations which have "to be expected" by the traffic participants including autonomous vehicles. The guidelines should define a minimal operational design domain, ODD.

Thus, this paper does not aim to give guidelines for

- completeness and robustness of technical solutions,
- suitability of user interfaces, etc.

These topics are related to the specific technical design solution. This should be left to the providers of such functions; competition should aim to develop the best solution within the design space. And it should not hinder anybody to provide solutions which even might expand the minimal ODD.

As already pointed out in (Schöner 2014), the risk of dangerous traffic situations derives from three factors (as for any other accident types), see Fig. 1:

---

[3]See Footnote 2.

[4]http://arxiv.org/abs/1708.

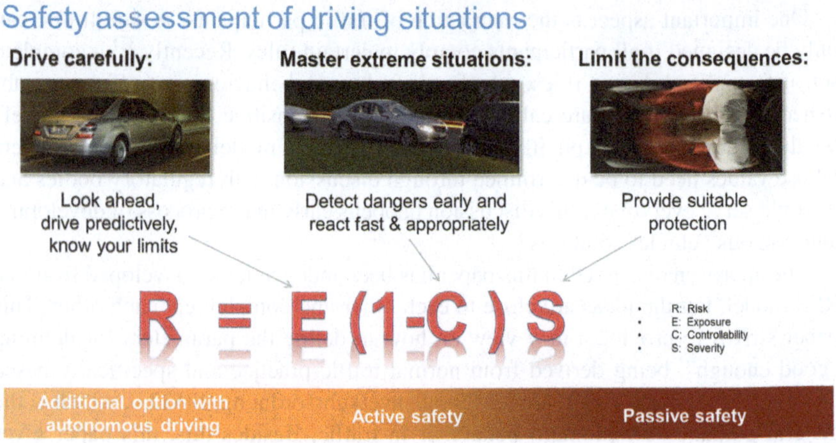

Fig. 1 Risk results from exposure to, controllability of and severity of accidents (Schöner 2014)

- S: severity of an accident, once it has happened
- C: controllability (avoidability of an accident), if an accident-prone situation comes up
- E: exposure to accident-prone situations.

Driver Assistance Systems DAS (up to level 2 of vehicle automation) have been developed in the past, in order to help the driver to control critical situations (bring the factor C closer to 100%). But they always left the task to the driver to avoid coming into safety critical situations (no influence on the factor E). Since in level 2 systems, most systems are designed to not interfere with the driver unnecessarily, the intervention of the DAS thus was limited in most cases to situations, when an accident is getting almost unavoidable. For this reason, there has been a big focus on very intense interventions (emergency braking) of the driver assistance system, striving to increase the controllability C of the situation. Autonomous vehicles, however, have a chance to perform automatic vehicle reactions in a much earlier state. This allows to intervene in a more comfortable way, which finally has the effect of reducing the exposure E to truly critical traffic situations. For autonomous vehicles of level 3 and higher, only reducing the exposure E to critical situations becomes the final reason for reducing the risk of accidents, since the controllability C should be already inherited from level 2 systems.

3 Limits of Controllability

Figure 2 shows the typical characteristics of controllability distributions in a critical situation (Schöner 2014): the red line shows the controllability of unassisted human drivers, the green line the curve of assisted human drivers, and the blue line the curve

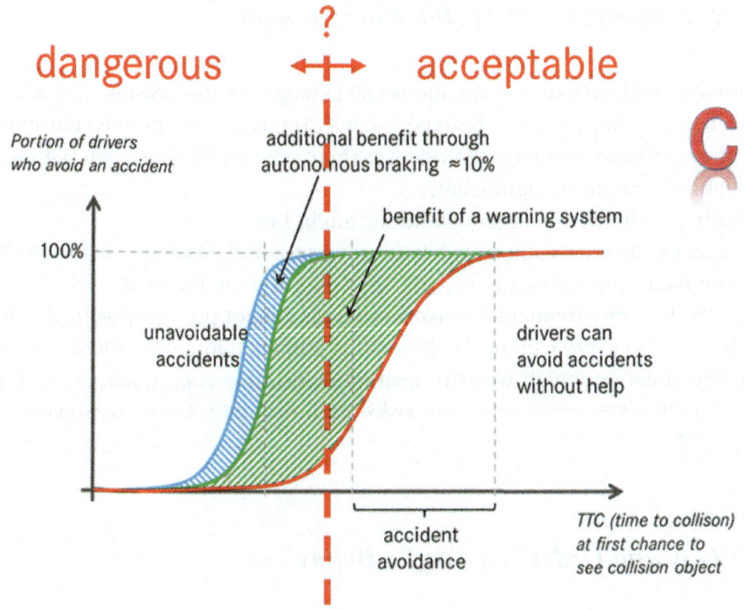

Fig. 2 Limits of controllability (Schöner 2014)

of automated collision avoidance systems. All three curves show, that there is a minimal reaction time which limits the controllability; typically, automated systems can react faster than human drivers, but they still cannot anticipate accidents. If the available reaction time is shorter than the necessary reaction time, controllability C goes down to zero because of physical reasons. In practical traffic this can be cargo falling off from a preceding vehicle, or an object being dropped from a bridge.

It must be accepted that there exist conceivable situations, in which a single traffic participant (no matter if human or AV) will not be able to avoid an accident, once a situation has evolved to a certain criticality. A situation which becomes close to being uncontrollable, is considered dangerous. An agreement on a value for a limit in the available reaction time helps to draw a line between "avoidable" and "unavoidable" (force majeure, natural disaster) accidents.

The only way to avoid accidents caused by such uncontrollable situations is, to make sure to never come into such critical situations (which is equivalent to limiting the exposure E). However, there are two main categories of situations which a single traffic participant cannot rule out by himself.

3.1 "Emergencies" in Traffic Environment

In real world, sudden extreme situations can emerge; for this reason, they are called "emergencies". They are not to be expected in normal daily operations (low exposure E), but they can have extreme consequences (high severity S); thus, their risk (without controllability) might be significant.

In traffic, such unusual outer conditions might be:

Unexpected severe weather, landslides, road and bridge failures, sudden technical vehicle faults, terroristic or suicidal actions, lost cargo on the road, …

Even the best environmental sensing methods cannot provide practical solutions for such risks. For every technical solution (as complex it might be) there is always an imaginable situation which will still lead to undesirable consequences. Commonly accepted methods to cope with such risks are a necessity for a "reasonable safety level" of AVs.

3.2 "Law and Order" in Traffic Behaviour

Traffic is governed by general laws and behaviour rules (order). If everybody would comply to all laws and rules, the behaviour of other traffic participants would be relatively easy.

Traffic laws are considered unconditionally valid. So, there is no question that autonomous vehicles have to adhere to the laws. And the AV planning software can assume in planning its own behaviour that every other traffic participant is adhering to the laws as well.

Traffic rules, however, regulate the *orderly* and thus *safe* behaviour of traffic. Normally, everybody should stick to those rules as well. But in practical traffic, the rules are often not obeyed completely: for practical reasons in order to solve unusual situations, to ensure traffic flow (e.g. if a lane is blocked, even a solid line as lane marking needs to be crossed), or just by loose interpretation of the rule (how to stick exactly to a speed limit), or by mistake. Thus, traffic rules cannot be used as limits for the assumed behaviour of other traffic participants.

But what kind of behaviour can you really expect from other traffic participants? If you always consider worst case behaviour, traffic probably would collapse by over-cautiously driving participants. Some roads are designed and operated specifically (freeways) in order to reduce the variance of behaviour to be expected by other traffic participants on these roads. Guidelines (derived from those traffic rules, and dependent on road classes) are needed for the design of interactive AV behaviour.

In German and Austrian jurisdiction[5]—for example—a concept of "trust" has been established, which defines certain risks as acceptable: a traffic participant can trust in the fact, that other traffic participants do obey all relevant rules—he is not obliged to consider any unusual endangering misbehaviour when deciding about his

[5]https://de.wikipedia.org/wiki/Vertrauensgrundsatz.

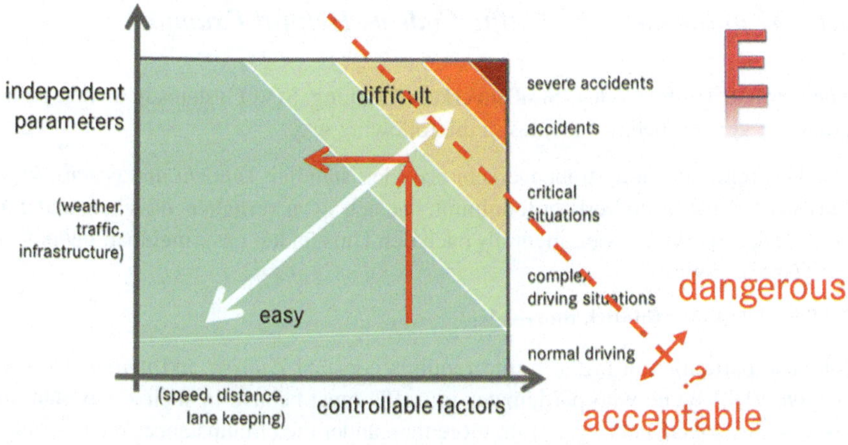

Fig. 3 Safe behaviour for avoiding dangerous situations (Schöner 2014)

own actions within the given rules. (However, if indications of misbehaviour of others are observable, he still must react by mitigating the risks from this misbehaviour.)

4 Safe Behaviour in Traffic

All over the world, driving behaviour in traffic is learnt when getting a driver's license. Safe behaviour is typically defined in very general abstract rules, but it is learnt intrinsically by practical training in traffic situations. Thus, in order to teach an AV a safe driving style, we cannot build on a given and complete set of exact rules.

Generally speaking, safe behaviour depends on many parameters (see Fig. 3): There are controllable factors, which can be influenced by the driver, namely: driving speed, longitudinal distance and lateral distance to other objects. And there are uncontrollable (independent) parameters, namely: weather conditions, traffic conditions, motion state of other objects, road curvature and lane markings, safety infrastructure along the road, and many more.

Driving safely means to use the controllable factors in order to stay away from critical situations (keep the exposure E small), which might emerge from complex driving situations; these factors have to be adopted constantly if easy situations turn to more difficult ones by changing outer conditions (see red arrows in Fig. 3). Safety depends also on limitations due to own driving abilities, which might be related to visibility, driving dynamics skills, tire conditions, etc. In summary, staying away from accidents and even severe accidents requires the recognition of a **level of danger** of the current situation, considering all those factors.

4.1 §1 of the German Traffic Code as Helpful Example

The German Traffic Code ("Straßenverkehrsordnung, StVO") states in its first paragraph the general behaviour rules in the following way:

"(1) Die Teilnahme am Straßenverkehr erfordert ständige *Vorsicht* und *gegenseitige Rücksicht*. (2) Wer am Verkehr teilnimmt, hat sich so zu verhalten, dass kein anderer *geschädigt*, *gefährdet* oder, mehr als nach den Umständen unvermeidbar, *behindert* oder *belästigt* wird."

Translated into English this reads:

"(1) The participation in the traffic requires constant *caution* and *mutual consideration*. (2) Anyone who participates in traffic must behave in such a way that no one else is *harmed*, *endangered* or, more than under the circumstances unavoidably, *hindered* or *bothered*."

These rules are the general basis (usually relevant in absence of specific traffic rules) for law suits and court judgements with respect to traffic accidents in Germany (König and Dauer 2013). With the help of some interpretation, specification and quantification these sentences can be transformed into more specific driving rules. They also destinguish between different behaviour categories, which are important for priority discussions.

4.2 Specification of the Key Words in the Traffic Code

Here are some specifications, how the key words used in the traffic code should be interpreted, or which aspects should be considered in this context:

Caution:	Looking far enough ahead, watch out for any signs of unusual or unexpected behaviour of other traffic participants, know your own skills and limitations, anticipate the consequences of your actions, <u>know</u> all the prevailing rules (driveable space, adequate speed, right of way, …), which implies to first know exactly where you are,
Mutual consideration:	Consider and anticipate behaviour of others, understand typical signalisation of others, drive predictably for other traffic participants, consider how your own behaviour is interpreted by others,
Harming:	Always avoid own accidents! (accident: undue touching of other participants or objects, leaving of driveable space) Do not cause accidents of others!

Endangering: Leave enough space & time for safety reactions,
do not cause situations with too short "time to collision" (TTC),
do not drive with too short "temporal headway" (THW),

Hindering: Leave enough space & time for comfortable reactions,
do not force others to accelerate/steer uncomfortably,
do not hinder normal traffic flow,
which implies to obey all prevailing traffic rules,
signal your braking and lane change intent clearly,
be cooperative (actively open gaps for safe distances to others)

Bothering: Do not stop in the way of others,
do not distract others,
avoid situations which may lead to increased level of danger.

These specifications are derived from normal traffic practice; of course, this is open to further specifications and additions. Most of these words refer to behaviour which aims at reducing the exposure to critical sitations, as explained in the context of Fig. 1. The lower four words from harming to bothering define an order of criticality in the sense of Fig. 3. The knowledge of prevailing rules and of the own performance limits (in category *caution*) is considered safety-critical; without correct knowledge a safe and situation-adaptive behaviour is impossible.

The "Responsibility-Sensitive Safety (RSS) model" defines five specific behaviour rules in the following way[6]; their relation to the above specifications is marked with the respective key word:

1. Do not hit someone from behind. (*harming*)
2. Do not cut-in recklessly. (*endangering*)
3. Right-of-way is given, not taken. (*mutual consideration, endangering, hindering*)
4. Be careful of areas with limited visibility (*caution, mutual consideration*)
5. If you can avoid an accident without causing another one, you must do it. (*harming*)

4.3 Quantification of the Key Words in the Traffic Code

The next step is to quantify the specifications of the key words. All values given here are considered as parameters for starting or promoting a concrete discussion. They need to be defined in projects like PEGASUS, or in similar international projects or standards resulting from such.

The quantification of the key words "endangered" and "hindered" is mainly based on human reaction times for safety-relevant and comfort-relevant actions in traffic

[6] See Footnote 4.

scenarios. Human reactions are generally limited by necessary perception, interpretation, reasoning and action times. The requirement for AVs must ensure equal or better total performance.

Human reaction times in detail depend on many factors, however they cannot be validated in every conceivable situation separately; thus, typical traffic reference situations must be agreed on. Typical human reaction times for such situations can be measured in simulator experiments; on the other hand, a conservative limit for human reaction time can be derived from existing Driver Assistance Systems (which produce warning sounds leaving reaction times which are statistically proven to be save enough). Such rough estimates should be good enough to be used for other situations as well, if no better knowledge is available. The definition of a limit for perceived "dangerous" situations and "hindering" situations can be verified by experiments as well.

Key word Quantification (suggested parameter values, only for start of a discussion!)

Caution: have a *quantitative* knowledge about own abilities and limitations:

know your own position, driving state and vehicle state,
know your preview horizon (from map) and your sensor range,
in summary: have "self-awareness";

know all rules and conditions on your route *redundantly*:

from an actively managed map (for preview)
and from sensors (interpreting traffic sign, lane markings etc. for final validation),
know variable signals (lane lights, traffic lights) *redundantly* by camera & Wifi;

have an *up-to-date* knowledge of track conditions and suitability for AVs:

location of emergencies and difficulties (objects on the road),
location of emergency vehicles and road works,
weather forecast;

define and apply a reasonable measure for a "level of danger".

Mutual consideration: signal your lane change intent clearly (3 s before),
signal unusual reactions clearly (heavy braking by flashing brakes),
be able to anticipate behaviour of others from their motion pattern,

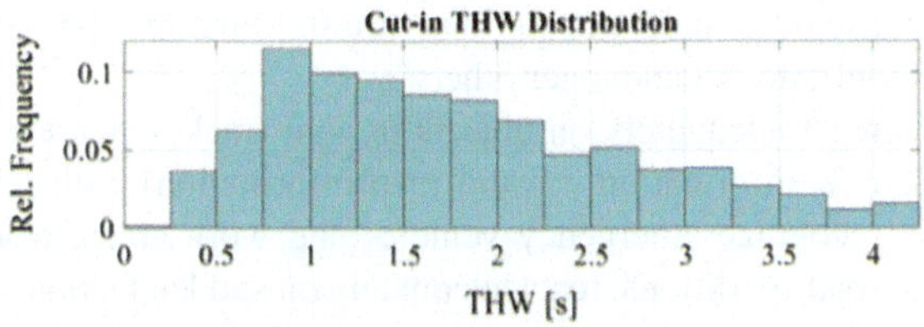

Fig. 4 Example for established driving discipline, here temporal headway in cut-in-situations on highways (Krajewski 2018)

be able to understand signalling (lane change, warning lamps, emergency vehicles);

Harming: No accidents *at all* driving alone,
No accidents *caused* (with participants) if TTC > 1.5 s;

Endangering: keep "time to collision" TTC > 1.5 s,
keep "temporal headway" THW > 0.7 s;

Hindering: obey speed limits (stay below +5%; assume others to stay below +20%),
stay in lane (with mirrors stay away 20 cm from lane markings),
do not force others to accelerate/steer uncomfortably $|a_{long}| < 2.5\text{m/s}^2$, $|a_{lat}| < 2.5\text{m/s}^2$,
be cooperative (open gaps for others so they can cut in with THW > 0.7 s),

Bothering: do not stop in the way of others (do not reduce the driving lane more than 10%),
do not drive unnecessarily slowly (less than 10% below traffic flow without reason).

Some quantifications can be derived from normal driving discipline in traffic measurements, as seen in Fig. 4: from the distribution of temporal headway in cut-in situations in normal highway driving a THW-limit of 0.7 s seems reasonable.

5 Corollary for *Single* Vehicle Traffic Behaviour

In order to sort the implications of the behaviour rules more clearly, let us concentrate in a first step on guidelines and performance tests for a single AV, without any other interfering traffic participant. This again can be divided into driving under normal conditions and under extraordinary conditions:

- Single vehicle, normal traffic situations; passing criteria "do not endanger": make sure you know all relevant rules for your location, obey the rules, stay in your lane, drive with comfortable dynamics, monitor your own system, stay away from own limitations, drive only in locations where your limitations comply with the requirements of the location.

- Single vehicle, extraordinary traffic situations; passing criteria "do not harm yourself, do not endanger others":
  react adequately on objects in your track, react adequately on surprises (sudden new information, invalid map information) with respect to trajectory planning, recognize emergency vehicles and work areas, react on changing weather and road conditions, react adequately on sudden technical failures, … and in all cases: do not react in a way which would endanger others or other objects.

5.1 Single Vehicle Performance Tests

Examples for performance tests which should be deducted to verify compliance with those guidelines are given below:

- Precise localisation of the vehicle, even under difficult conditions (no or bad GPS satellite connection)

  → know your position and deduct the prevailing rules and conditions for this location correctly; show adequate behaviour when localisation degrades.

- Safe and comfortable driving, even under worst case street curvatures for a given road class

  → Do not drive faster or closer than you have under control, w.r.t. speed—distance—lateral distance, without heavy accelerations longitudinally and laterally.

- Awareness of information horizon (sensors, map), ability to react on reduced horizon (sensor deficiencies due to weather or lighting conditions, loss of or wrong map information, loss of communication, denied road clearance)

  → Know your own look ahead and safe stopping performance limitations (safety and comfort limits of acceleration, deceleration, curve speed, sensing horizon) depending on weather and lighting conditions, and show adequate reaction based on this information.
  → Sensing horizon at high speed requires knowledge from a map, with sufficiently recent information—show adequate behaviour when the quality of map information degrades (outdated or missing map data, loss of road clearance)

- Ability to react on objects on the road (lost cargo: a definition of relevant objects to be detected is needed) or unexpected lane deviations (road works, potholes, …)

  → show your detection ability and adequate behaviour at highest speed and large sensing horizon
  → show that behaviour changes due to awareness of reduced sensing horizon is effective
  → If you do not know for sure (for example, redundant information is inconsistent)—assume the worst, show adequate behaviour depending on road type
  → prove that you have a (sufficiently) safe method to handle worst case scenarios!

Component and system robustness are not considered here in detail. Such design dependant reliability tests and verification of fall-back solutions should be covered under ISO 26262.

5.2 Autonomous Driving Starts with Autonomous Stopping

Bringing the vehicle into a safe state out of any situation is one of the most challenging performance requirements for AVs. An autonomous vehicle should not drive without making sure that it can safely stop the vehicle, even under worst case conditions. Even if there is a safety driver in the vehicle, safety concepts should require that the vehicle is able to stop without a take-over of the driver (there might be less stringent worst-case conditions for level 3 than for level 4 or level 5 AVs).

Stopping requires bringing the vehicle down to zero speed from a given initial speed within the distance, which is physically needed for deceleration. This distance defines the *necessary decision horizon* which is the look ahead perception horizon minus vehicle reaction distance (distance required for evaluation of the free space incl. safe trajectory planning and initiation of the deceleration). The necessary decision horizon depends mainly on the possible or accepted value of deceleration. For normal driving without urgent need of bringing the vehicle to a stop, a slower deceleration (goal: comfortable, and do not hinder others) should be considered than for emergency braking (do not endanger others). Scheduling of take-overs in AVs of level 3 at highway speeds should normally be based on map information, since sensor information would not provide a sufficiently large *available decision horizon*.

Thus, the map-based available decision horizon must be large enough to allow for

- stopping with comfortable decelerations (not *hindering* anybody), or
- stopping after a "non-responded take-over request" with a manageable deceleration cascade for followers (not *endangering* anybody)

The sensor based available decision horizon must be large enough to allow for

- autonomous stopping with a manageable deceleration cascade for followers (not *endangering* anybody)
- assuming tbd. minimum-size critical objects to be detected by the sensor set

The maximum acceptable speed can be derived from these conditions, in dependence on the availability of reliable map information. Test scenarios, in order to verify the ability to bring the vehicle to a full stop (with the adequate braking procedure), or in order to verify the correct reaction on the loss of reliable map information, can be derived from these considerations. In level 3, stopping in the lane should be acceptable, since eventually a driver can take over; level 4 and 5 might require finding a suitable stopping point without hindering others (if possible).

5.3 Blind Stopping Procedure

Sudden sensor degradation, or even a complete sensor set failure, might be caused by outer conditions (as well as by technical faults). No failure rate characteristics are (to my knowledge) available to date which allow to claim, that the sensing system will be available with very low failure rates for continuous AVs operation under any outer condition. Even robust radar systems, as well as any optical system, can have severe performance losses in very short time during extreme rain or heavy snow fall.

As long as no other evidence can be provided by an AV manufacturer, it is suggested that the vehicle performance in this worst-case scenario should be guaranteed and tested with the "Blind Stopping Procedure" as follows:

In addition to the normal driving trajectory, the vehicle should continuously generate a planned trajectory for stopping using a manageable deceleration cascade for followers (not endangering anybody), based on available map and sensor information. In case of a complete sensor failure, the vehicle should be able to switch to this trajectory based on last available valid sensing data plus inertial and wheel data (by ESP/ESC control); the vehicle should be able to come to a full stop without leaving the driveable surface of the road ahead. If sensing system and ESP/ESC-system are well designed (i.e. have independent power supplies), this concept would also help in designing for high technical system reliability. Such an emergency stop should be clearly indicated by warning lamps and V2X communication.

5.4 AV Highway-Pilot: Road Clearance and Map Learning

A highway ("Freeway" or "Autobahn", with separated lanes for different driving directions and restricted access for bicycles and pedestrians) is a designated, specifically protected and constantly managed traffic area. Here vehicles can drive with less safety margins than on other roads because of these conditions: wider lanes, normally no risk of oncoming traffic, normally no unaware pedestrians, etc.

For the AV concept of "Highway Pilot" as considered in the PEGASUS project, a clearance for autonomous driving has to be obtained from a central server for using autonomous mode. The clearance is related to a dated map of the road section. The information in this map can be assumed valid for planning purposes (to ensure comfortable reactions), however deviations from the map must be considered and map information has to be confirmed by vehicle sensors (to ensure safety reactions). Any safety relevant deviation from the map information needs to be reported back to the central server, in order to provide a constantly learning map. The map information is provided in several layers (Bock 2018), and it includes:

- road layout: road network, lanes and their interconnections, driveable surfaces, …
- traffic rules: traffic signs, lane markings, locations of variable signals, …
- road condition: moisture, snow and ice coverage, friction coefficient, …

- traffic condition: congestion, accidents, debris, emergency vehicles, road works, ...

Especially, the last two categories can change in a very short time, and they might be essential for safe behaviour.

One essential parameter for the safety assessment is the update speed of the map and the information delivery rate to the AVs. It is impractical to require that a complete collision avoidance with fast changing objects based on this map information can be realized. However, the performance of this crowd-based learning needs to be specified in order to define an adequate level of contribution to the safety concept.

Examples for conditions for a release of such a system:

- The clearance for AVs must be reissued/reconfirmed at least every xx minutes.
- Every AV must participate in a feedback loop as condition for AV clearance.
- If several independent background servers are available, safety critical information must be exchanged to each other within an update rate of xx minutes.

For a fast building-up of a reliable learning map network, it should be considered that even level 2 vehicles could significantly contribute in the feedback loop, at least for some of the information categories.

6 Corollary for Behaviour in Interactive Situations

Based on the fact, that *everybody* should comply with the rules from the traffic code, certain behaviour of traffic *partners* can be expected in *any* interactive traffic situation. In the interpretation of the traffic code in König and Dauer (2013) this is called "trust" in the safe behaviour of others.

- Since an AV should not be harmed or endangered by any other traffic participants, very low TTC or THW are not to be expected for its own planning purposes (some extraordinary scenarios might be specified with extreme values);
- Managed highways could have a different limit for certain to be expected behaviour than other roads (needs to be specified with the street class).

On the other side, a safe and robust reaction on (rare) expectable situations (collision free, if only being hindered or bothered by other participants) needs to be ensured. Reference situations for such behaviour should be defined, based on the general behaviour rules. In general, for city driving with all kind of different traffic participants, it is a large task to iterate through all possible interaction types. However, for the Highway Pilot this leads to a limited number of possible interactions; this should be the focus in Sect. 6.3.

6.1 Testing of Borderline Interactive Situations

As explained in Chap. 3, interactive scenarios will lead to uncontrollable situations when criticality parameters are increased. Shorter available reaction time will always decrease the ability to control a situation. This can be caused by shorter initial interaction distances in the beginning of a critical scene, or by higher relative speed, or by higher acceleration/deceleration. Since there is a monotony on the dependence of such parameters, borderline situations will guarantee controllability on one side of the borderline case in this parameter space.

The principle for testing for safe behaviour in interactive scenarios should be divided into two cases:

- Interactive scenario, <u>normal</u> traffic situations; pass criteria "do not endanger":
  Define traffic situations, in which one participant needs to perform a challenging task in traffic, which might involve getting into the intended path of other participants. The *first* participant needs to show in the borderline case that he would <u>not</u> act in a way which would endanger others. For scenarios less critical than the borderline case, the first participant could—without blaming him—perform his manoeuvre.
- Interactive scenario, <u>extraordinary</u> traffic situations; pass criteria "do not harm" (collision avoidance):
  Define traffic situations, in which one participant behaves at the lower borderline limit to endangering a second participant. The second participant needs to show controllability in this situation (i.e. without collision); but for more endangering or even harming behaviour of the first, the second participant may not be liable if he cannot avoid an accident. More than borderline dangerous behaviour of the first participant *may be* compensated by the second participant, but it *cannot be guaranteed.*

6.2 Safety Impact of Behaviour Interaction in Traffic

Using the concept of behaviour within the categories of the traffic code, several combinations of outcome of the interactive situations are possible (see Fig. 5).

Interaction of participants, acting both with normal, bothering or even hindering behaviour should not lead to any accidents. If one participant acts in an endangering way, accidents may be possible, even if the other participant acts normally, or bothering, or even hindering. That means, there are safety margins which can be considered as "forgiving behaviour" for endangering behaviour of the other participant. Even if both participants act in an endangering way, the situation can end without any accident, although this is less probable. Harming action by any participant leads to an accident, by definition.

The visualisation of this concept also shows, that it would be not safe to limit the behaviour of all participants to "not harming" for the borderline cases. There would

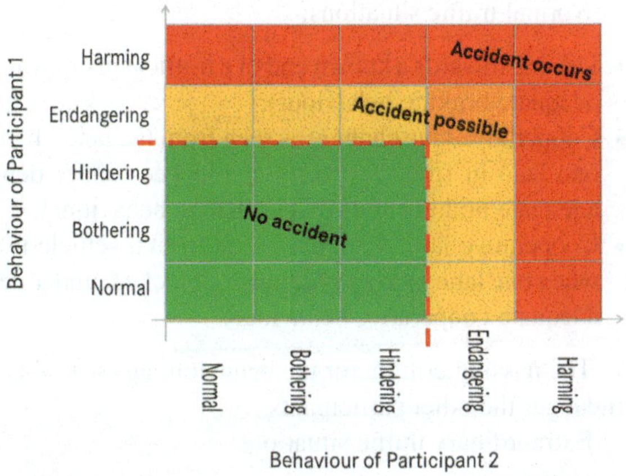

Fig. 5 Logic of traffic safety: dependance on behaviour interaction

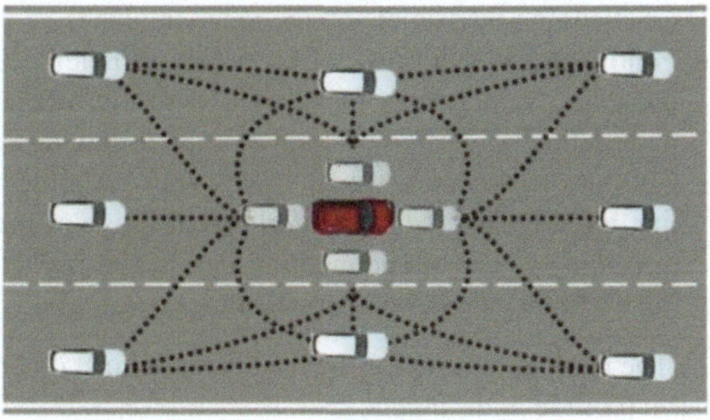

Fig. 6 Complete set of two-vehicle highway interaction scenarios (Bock 2018)

be no margin for small variations in the behaviour. The region of "not endangering" others is important to avoid accidents with a certain robustness.

6.3 *Defining Borderline Cases for the Function "Highway Pilot"*

In Bock (2018) a complete set of interaction scenarios with two vehicles for possible collisions on highways is presented, see Fig. 6. It shows 16 different ways which can lead to longitudinal or lateral collisions, resulting from different initial constellations. In partial overlap with this, in Schöner (2018) the three different basic scenarios of following a preceding vehicle, of cut-in and cut-out manoeuvres with sudden view on previously invisible third vehicles/objects on the road are described as challenging scenarios.

For all these different scenarios, borderline cases at the limit of *endangering* behaviour (of the causing traffic participant) can be constructed. These include the following most important examples:

Normal traffic situations:

- Driving towards a known end of a traffic jam or a known object on the road (display adequate braking behaviour);
- Cooperative lane change, as seen from the lane-changing vehicle, especially when one lane in front is ending or blocked, under dense traffic conditions (display adequate braking and/or lane change behaviour).
- Cooperative lane change, as seen from a vehicle in the receiving lane, especially when one lane in front is ending or blocked, under dense traffic conditions (display adequate cooperative behaviour).

The passing criteria for the behaviour against other participants should be, not to endanger the other participants.

Extraordinary traffic situations:

- Cut-in of a vehicle with high differential speed and/or heavy braking in front (challenging the vehicle on the receiving lane).
- Following and full brake of preceding vehicle because of a blocked road ahead; preceding vehicle may be stopped violently by crashing into the end of a traffic jam (define borderline challenge of the following vehicle).
- Cut-out of preceding vehicle gives view on a blocked road (or even an oncoming third vehicle; define borderline challenge of the following vehicle); since the oncoming vehicle is endangering (and heavily against the rules), an outcome without accident cannot be guaranteed.

The passing criteria for the behaviour against other participants should be, not to harm (collide with) others.

7 Learning of Safe Driving

Human drivers develop their full driving skills only after some time of driving experience. We accept that new unexperienced drivers start driving in public traffic without this driving experience. So, the chance of making little mistakes or needing longer reaction times than experienced drivers is higher than for the average of the driver community. Designing the traffic behaviour with reserves for such unexperienced behaviour is mandatory to allow for robust traffic management under such conditions. This way, small mistakes will not lead necessarily to a collision, but a resulting endangering situation might be mitigated by the forgiving (robust) behaviour of others.

The acceptance of unexperienced behaviour only works for a small number of traffic participants. Gaining experience over time by learning mechanisms is essential to make this work in the whole community. We expect from and trust in human drivers that they learn more and more safe behaviour during their first years of driving.

Transferring this concept to the introduction of AVs requires to establish learning mechanisms for AV driving software as well. The industrial concept however is to introduce AVs after an intense development phase. During this phase, feedback loops for learning safe behaviour are built into the integrated development and testing processes of the vehicle developers. Even after the release of AVs, the monitoring of vehicle behaviour and constant improving of the product need to go on.

Let us have a closer look on how the learning of driving skills works (Schöner 2018).

7.1 Human Driver

When a driving examiner assesses the driving skills of an applicant for a driver's license, he considers the driver's behaviour with respect to his judgement of the entire driving situation including

- own skills of driving
- conditions and abilities of vehicle
- own position and driving state, traffic, weather
- recognition of prevailing rules, their applicability and limitations
- checking of evolving driving situations in compliance with intention and plan
- ability to update a navigation plan in due time
- handling of surprises
- correct assessment of own (and other's) reaction time.

Generally, in driving examinations, no assessment of collision avoidance abilities of the applicant is verified. However, the examinant assesses by his experience, whether the applicant has a sufficiently well-established (implicit) set of behaviour rules based on general values from training in a driving school.

Besides this, in many countries driving with an experienced driver (supervised driving) has been established as an additional measure for fast improvement of the driving skills of young drivers. Even insurance rates are reduced for young drivers using this method.

What happens during the supervised driving phase is as well not focused on improving collision avoidance skills of the driver but improving the mitigation of dangerous situations (reducing the exposure to collision-prone situations). The supervisor helps the unexperienced driver by focusing his attention on early indicators of possibly risky situations and behaviour (of the driver or of others). Especially the knowledge of the own limitations (for example: with respect to fast and correct judging of situations; implications of the weather conditions on the own abilities) are an essential part of the support of the supervisor. Again, there is no clear algorithm on how to recognize such situations; the intuitive skills of the human supervisor are essential for the effectiveness of this learning procedure.

Fig. 7 Possible "safe driving score" for human drivers (Schöner 2018)

7.2 *"Safe Driving Score" for Human Drivers*

In order to copy this successful human strategy for accident reduction to autonomous vehicles, the main challenge is to assess complexity and criticality of the situation ("level of danger") and judgement of the own capabilities ("self-awareness") based on adequate measures. With these goals, the definition of measures for the safety of driving is essential. This must include also low criticality levels, not only the well-established measures for collision prone situations (using time-to-collision, TTC).

For a human driver, the driving style could be assessed, for example, by giving him feed-back on key parameters like distance, speed and look-ahead time (see Fig. 7)—as would probably a supervisor do. Even a fuzzy algorithm, which would judge these parameters based on simple assessment methods, would give the driver a feed-back, which he intuitively understands. The driver would learn a safer behaviour; this should work like driving with better fuel economy based on similar fuzzy indications. A "safe driving score" like this one could (at least partially) do a similar job compared to a human supervisor.

Keeping a safe distance or a safe speed depends a lot on outer conditions. The driver needs to learn, under which conditions he gets good or worse scores, even with the same driving style, when using it in different situations.

It is hard to implement an assessment of the own driving skills (self-awareness) in a score for the driver. This must be left to the driver himself.

7.3 *Self-awareness of Autonomous Vehicles*

For an autonomous vehicle, the driving style is programmed into the vehicle; it makes more sense to assess the robustness of autonomous driving with respect to several categories of outer driving conditions. These categories could be the map quality, the visibility of tracks (lane markings), the traffic conditions, or the weather conditions (see Fig. 8).

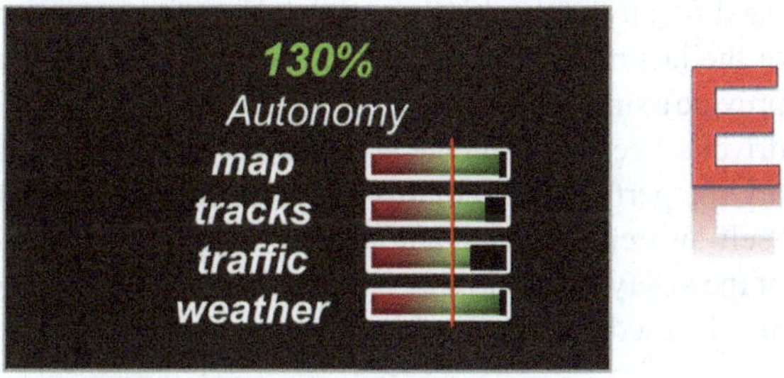

Fig. 8 Possible "safe driving score" for autonomous vehicles (Schöner 2018)

It should provide insight into the question:

How robust *was* the safety of autonomous driving?

For scores above 100%, the vehicle would be able to adopt the driving style accordingly. If one of the categories would fall significantly under 100%, this would be a reason to hand back the driving task to the driver (in level 3 vehicles) or to suspend the driving completely (for level 4 and 5 vehicles).

In detail: looking back for the last driving minutes, the *a posteriori* assessment of driving segments could cover the following topics (and explain, why the vehicle changed to a more defensive driving style with respect to the → indicated aspects):

<u>**Map**</u>

- was the map w.r.t. drivable surface, traffic signs, traffic jams in accordance to the experienced reality and up-to-date, were there any unexpected objects or surprises, any wrong map information?
  → adequate speed and distance to others, look ahead distance

<u>**Tracks**</u>

- Were the tracks well visible, were they marked as expected?
  → adequate speed

<u>**Traffic**</u>

- Did the traffic behave in a predictable way, were there any dangerous objects or driving styles, dangerous cut-ins or partially blocked lanes?
  → adequate time gap and longitudinal & lateral spacing

<u>**Weather**</u>

- Were the sensors able to cope with the weather, was the weather forecast correct?
  → adequate speed, distance, time gap

These parameters provide fuzzy, but evaluable measures for a "safe driving score", either continuously updated, or in summary after a complete ride. They can support the developer in monitoring the performance of the AV driving software, and to a certificatory it provides a measure for judging, whether the AV has a correct sense of the difficulty of the driving task it just performed. Comparing the human assessment of the just monitored driving task with the scores generated by the machine would provide a similar trust in the performance of the driving software as inquiring a human driver's-license-candidate at the end of the test drive on his impressions on his own driving performance. In this way, this type of "safe driving score" is a measure for "self-awareness" of an autonomous vehicle. It provides evidence for its perception of the safety of the driving situation and the way it is responding to it. The score can also be used for Artificial Intelligence learning and improvement of safe driving.

7.4 Implementation and Evaluation

How can such a score of self-awareness be implemented?

The path planner in any AV controller should have such measures already available, because the choice of the best trajectory and the suitable speed needs basically the same rating. However, such performance measures would normally not be available for vehicle validation. At least for certification purposes, but perhaps also as a monitoring feature to increase user trust it should make sense to make these measures available.

The path planner normally would not look back, in case of a surprise or other information only available at the end of a driving scene. So, the scores of the path planner probably would need some extensions to cover the full possible scope.

How can it be evaluated?

Short and long-time evaluations of the scores can make sense:

- Checking of parameters after every critical or even complex situation, whether the vehicle has realized its criticality (checking the self-awareness of the vehicle).
- Monitoring the evaluation of parameters over time to check the learning of the vehicle (verifying the learning ability of the vehicle).

A verification of the ability of the vehicle to perform such self-awareness should be considered as an important part of an AV driver's license (or certification procedure). The autonomous vehicle should this way make evident that it is able to judge its own capabilities and limitations, and that it is able to stay safely away from these limitations under any driving condition.

8 Summary

This paper describes a concept for defining "Good Enough" behaviour of AVs in challenging traffic situations. It starts by deriving a consistent set of principal behaviour rules from the behaviour guidelines in §1 of German Traffic Code. The exact parameters of this approach still need to be quantified by experts, to be derived for example from human behaviour in traffic, in situations which show safety by "*proven in practice*".

The behaviour in normal situations should be oriented on the parameters for "not *endangering*" others, instead of on "*no collision*" (as is standard for level 2 systems). This gives an extra safety margin necessary for the distribution of behaviour to be found in real traffic, like deviations from "acting to the rules". On the other hand, it releases responsibility for collisions in situations where endangering behaviour of *other* traffic participants caused the accident.

Borderline test situations can be derived from the quantified behaviour rules as reference situations for demonstrating safety for certification. Worst case test conditions are proposed for extreme cases which are needed to be defined by the community, not by the single manufacturer.

These rules should be good enough for introduction of AVs, however the vehicles should be able to show that they implement a sense for "level of danger" and "self-awareness", for assessing the performance and effects of AVs as traffic participants. Based on these measures, learning and improvement over time can be implemented.

References

Bock J et al (2018) Data basis for scenario-based validation of HAD on highways. In: 27th Aachen colloquium automobile and engine technology

König P, Dauer P (2013) Straßenverkehrsrecht. Beck'sche Kurz-Kommentare, Band, vol 5, pp 422–424; Verlag C.H.Beck, München

Krajewski R et al (2018) The high dataset: a drone dataset of naturalistic vehicle trajectories on German highways for validation of highly automated driving systems. In 21st IEEE international conference on intelligent transportation systems, IEEE

Schöner HP (2014) Challenges and approaches for testing of highly automated vehicles. In: 3rd CESA automotive electronics congress, Paris

Schöner HP (2018) Simulation for development and testing of autonomous vehicles. In: 18th Stuttgart international symposium, Stuttgart, 2018-03-14

Schöner H-P (2018) Self-awareness of autonomous vehicles. In: Symposium on testing and certification of autonomous vehicles, Nanyang Technological University, Singapore, 2018-09-14

Steininger U, Schöner HP, Schiementz M (2015) Requirements on tools for assessment and validation of assisted and automated driving systems, vol 7. Tagung Fahrerassistenz, München

Dr. Schöner is an independent consultant (and CEO) of "Insight from Outside"—Consulting. In addition, he is active as a member of the Driving Simulation Association, DSA. In 2018, he retired from his development position at Daimler AG in Sindelfingen. Until March 2018 he was Senior Manager "Testing Concepts and Test Site" (RD/ASK) in Daimler's R&D Center "Assistance Systems, Active Safety and Testing". In this and former positions, he was responsible for development and supply of methods for testing and validation for future chassis and assistance systems, including autonomous driving functions. This included methods to provide reproducible testing situations on proving grounds with automatically driven coordinated vehicles, simulation methods and proving facilities for function development and testing, and operation of the Driving Simulation Center of Daimler AG. Dr. Schöner played a key role in defining the German government-funded research project PEGASUS, which has the goal to specify "How good is good enough? And how can we prove this?" with respect to the driving task of AV, http://www.pegasus-projekt.info/en/home.

DENSE: Environment Perception in Bad Weather—First Results

Werner Ritter, Mario Bijelic, Tobias Gruber, Matti Kutila and Hanno Holzhüter

Abstract This paper presents the first results of the publicly funded ECSEL project DENSE (Adverse weather environmental sensing system). DENSE seeks to eliminate one of the most pressing problems of automated driving: the inability of current systems to sense their surroundings under all weather conditions. The task in DENSE is to develop a sensor suite for automatic driving, by means of which the vehicle environment can be reliably detected 24/7 under these bad weather conditions. In this paper, the state of the art of environmental sensor technology is first examined and evaluated in the CEREMA weather chamber. Then, the architecture of the DENSE Sensor Suite is presented and the development results of the most important system components are described. The results show that the realization of a 24/7 all-weather sensor suite is absolutely feasible with these components .

Keywords ADAS (Advanced driver assistance system) · Automatic driving in bad weather · Autonomous driving in bad weather · Inclement weather · 24/7 All-weather sensor suite · Short wave infrared gated camera · Gated camera · Short wave infrared lidar · SWIR Lidar · Pixel-accurate depth map · Signal enhancement · Robust learning technique · Defogging · Deep neural networks for signal enhancement · Global adversarial network (GAN) for signal enhancement

1 Introduction

Safe autonomous driving at level five requires a perfect environmental perception in all conditions, even in bad weather (fog, heavy rain or snow). The problem is

W. Ritter (✉) · M. Bijelic · T. Gruber
Daimler AG, Wilhelm-Runge-Str. 11, 89081 Ulm, Germany
e-mail: werner.r.ritter@daimler.com

M. Kutila
VTT Technical Research Centre of Finland Ltd., Tampere, Finland

H. Holzhüter
Ibeo Automotive Systems GmbH, Hamburg, Germany

J. Langheim (ed.), *Electronics Components and Systems for Automotive Applications*, Lecture Notes in Mobility,
https://doi.org/10.1007/978-3-030-14156-1_12

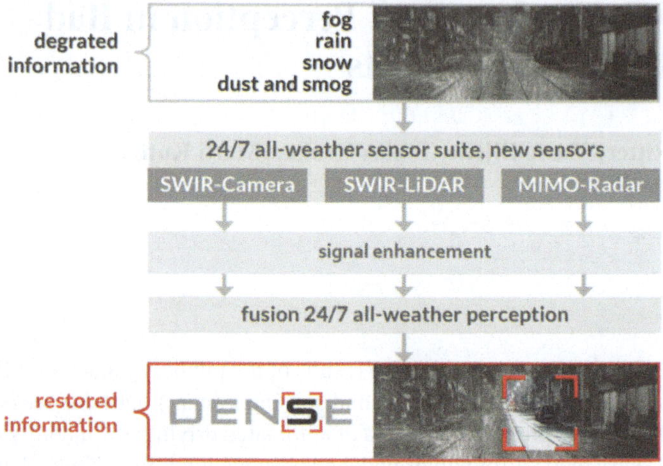

Fig. 1 DENSE system architecture

that today's sensors and the subsequent perception processing units are not designed for bad weather conditions (Källhammer and Cassel 2017). The sensors partly lose considerable performance in bad weather and the subsequent processing unit cannot adequately handle these performance losses or even compensate for these. To cope with this challenge advanced sensors technologies and new perception processing scheme, robust to disturbances, are needed.

The publicly funded EU project DENSE (see www.dense247.eu) addresses this problem. The objective of the project DENSE is to develop and validate a sensor suite for driver assistance and automated driving **working 24/7—in all weather.** The key components for this sensor suite are three new sensors, namely gated Short-Wave-Infra-Red-camera (SWIR-camera), SWIR-Lidar and a Multiple-Input-Multiple-Output-radar (MIMO-radar), and a novel processing unit consisting of deep neural networks to enhance and fuse the sensor signals. Based on the deep neural network the disturbed input signals could be possibly restored. The output of the sensor suite is an enriched and enhanced multi spectral image allowing to drive all ADAS functionalities also under bad weather conditions (Fig. 1).

The project started in June 2016. In this paper the first results of the sensor development as well as of the processing unit are presented.

2 State of the Art

Currently, as state of the art sensors, mainly passive CMOS cameras, 905 nm Lidars and low resolution radars are used for environment perception task of ADAS. They work well under normal weather conditions. In case of adverse weather, there arise

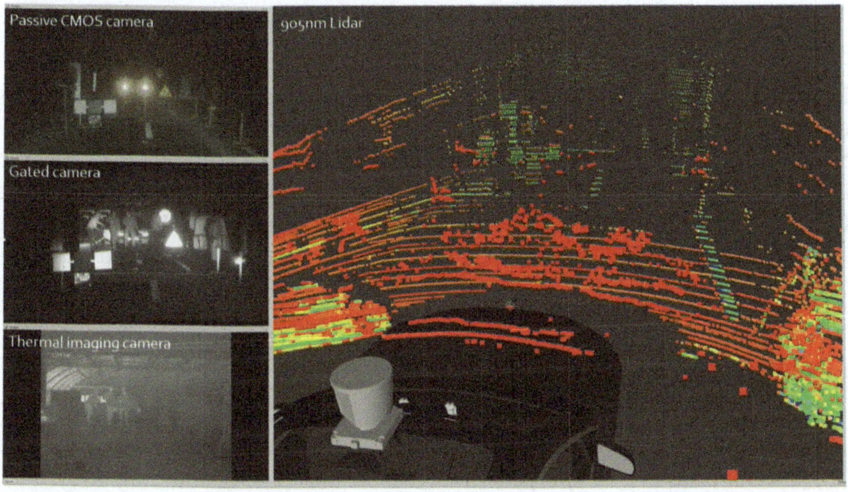

Fig. 2 Result images of four different sensors from the fog chamber with a visibility of 50–60 m. "Standard sensors" degrade strongly in fog, only the gated camera still delivers good images

a lot of problems with these state of the art sensors. Test results in the CEREMA fog-chamber in Clermont-Ferrand (Colomb et al. 2008) have shown that "standard sensors" degrade strongly in bad weather and especially in dense fog. The result images from Fig. 2 obtained by measurements in a fog chamber with a visibility around 50–60 m, illustrate this:

- Gated Camera: Both pedestrians and infrastructure are clearly visible in the image.
- Passive CMOS camera: The pedestrians near the oncoming vehicle with the headlights on are not visible.
- Thermal camera: The pedestrians are partly visible, but large parts of the infrastructure are invisible.
- 905 nm Lidar: The sensor detects a wall of fog. The fog wall around the sensor absorbs almost all energy of the Lidar. Objects further away are only weak or no longer visible.

This makes it clear that these sensors cannot be used for a driver assistance system that is supposed to function also in bad weather. New sensors, which also function under these conditions, must take over this task. In DENSE, two new sensors are being developed for this purpose with a gated SWIR camera and a SWIR Lidar, which are capable of sensing their environment sufficiently well even under these adverse environmental conditions. If disturbances still remain in the sensor signal due to poor environmental conditions, a downstream processing unit consisting of deep neural networks should compensate them as far as possible. The newly developed sensors and the deep neural network for signal enhancement are explained in more detail in the following sections.

3 Lidar

From today's point of view, the Lidar sensor is indispensable for fully autonomous driving according to SAE J3016 standard level 5, due to high safety requirements. SAE level 5 in this context means that the driving function must still function even in bad weather. For this reason, we have conducted initial performance tests in the CEREMA fog chamber (Colomb et al. 2008) with currently available state of the art Lidar systems, which usually operate at 905 nm. These were the Velodyne HDL-64S2, Velodyne HDL-64S3, Ibeo LUX and Ibeo LUX HD. A technical summary of these models can be found in Table 1: Technical summary of the tested laser scanners. Table 1. A more detailed description of the performance test can be found in (Bijelic et al. 2018).

The test was carried out in a test setup as outlined in Fig. 3 and shown in Fig. 4. The test scene consist of characteristic target objects such as an old exhaust pipe (0), reflector posts (1), various pedestrian mannequins (2,3,4), traffic signs (5), a tire without rim (6), a car (CAR2), road marking and multiple reflective targets. The Lidar sensors are located at the beginning of the chamber in which the test vehicle (CAR1) is located. Target objects in the test arrangement were an old exhaust (0), reflector posts (1), various pedestrian mannequins (2,3,4), traffic signs (5), a tire without rim (6), a car (CAR2), road markings and multiple reflecting targets. The Lidar sensors were located at the beginning of the chamber where the test vehicle (CAR1) was located.

We performed the test with artificially generated advection fog with different visual ranges V ranges from full visibility until a visibility of 13 m. Figure 5 shows

Table 1 Technical summary of the tested laser scanners

Manufacturer model	Velodyne HDL-64S2/S3D	Ibeo LUX/LUX HD
Laser layers	64	4
Range [m]	50/120	50/200 30/120
Multiple echoes	1 2	3
Heavy duty	No	No yes
Wavelength [nm]	905	905
Data rate [Hz]	5–20	12.5/25/50
Horiz. FOV [°]	360	85
Horiz. resolution [°]	0.0864–0.1728	0.125/0.25/0.5
Veil FOV	26.33	3.2
Vert spacing	1/2 to 1/3°	0.8
Embedded software	No	Yes
Distance accuracy [cm]	$\leq 2@1\sigma$	4
Operating temp. [°C]	−10 to 50 60	−40 to 85
Points per second	$\approx 1.3 \times 10^6 \approx 2 \times 10^6$	$\approx 44 \times 10^3$

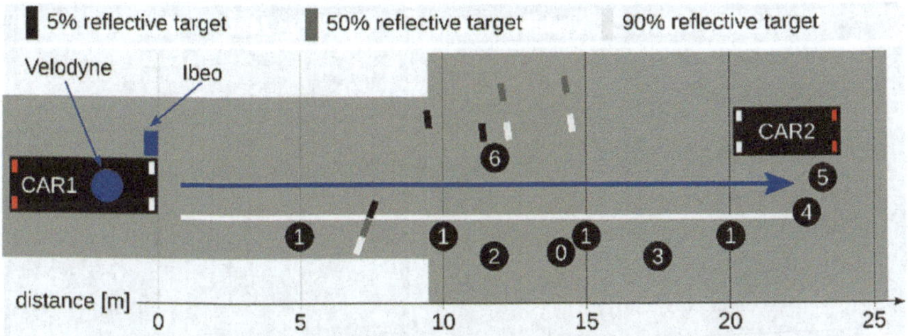

Fig. 3 Schematic drawing of the experimental setup in the fog chamber

Fig. 4 Experiment setup in the fog chamber

the Lidar birds eye view gathered by Velodyne HDL64 S2 laser scanner, which is typically for all 905 nm Lidars. This figure illustrates the heavy point cloud degeneration due to fog. With increased fog density, more laser pulse power is attenuated by fog. This leads to early and false fog backscatter. Therefore the sensors loses important true object echoes and leads to a reduced perception range. Furthermore the false backscatter creates clusters of false points leading to "fog walls", which can cover whole objects in close vicinity. Worst performance is achieved at a visibility of 13 m, where only the fog wall is detected by the sensor. The rest of the scene is invisible.

Other Lidar sensors operating in the nearfield-infrared (NIR) spectrum (at 905 nm), such as the IBEO Lux sensors, show a similar performance drop (see maximum viewing distance depicted in Fig. 6). This is due to the fact that the Lidar systems within the NIR band are limited in their maximum power so that they still reach laser class 1, which is considered as eye safe.

The above-mentioned problems caused by the laser intensity limited by eye safety become even more noticeable if the performance data of a Lidar system is to meet the system requirements of a SAE level 5 system, which means that the resolution and

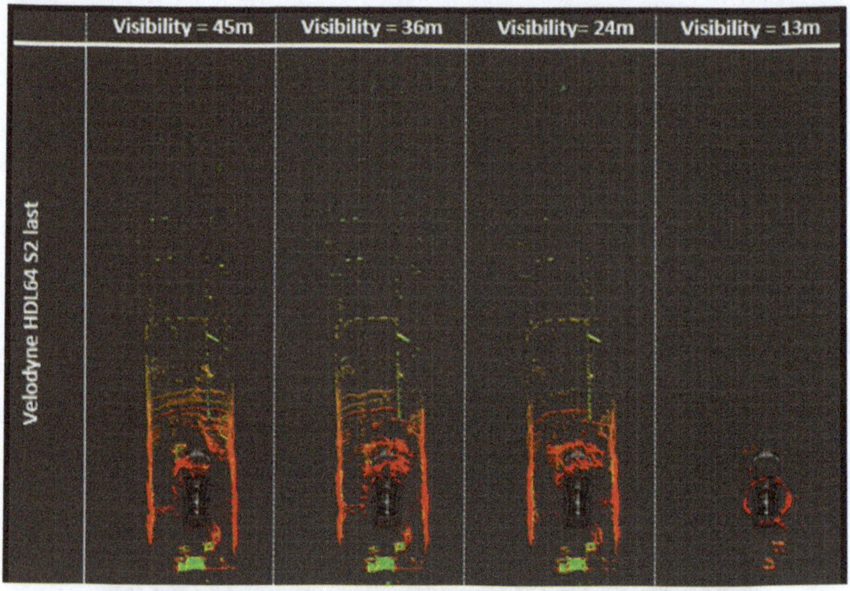

Fig. 5 Bird view of the Velodyne sensor HDLS64 S2 for different fog types and fog densities. Color encodes the intensity

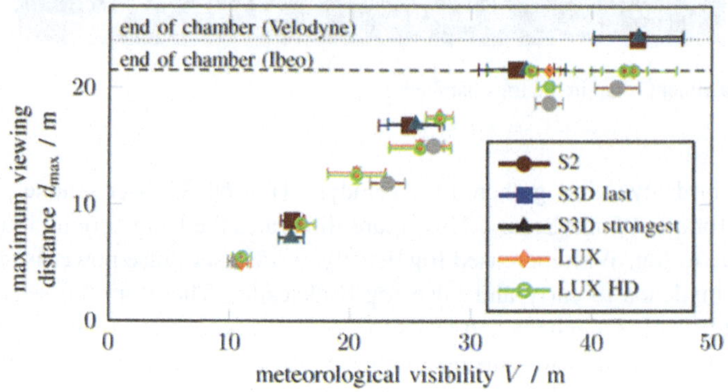

Fig. 6 Maximum viewing distance comparison for all sensors in advection fog and reflective targets with 90% reflectivity

range of the Lidar systems must be significantly increased compared to the existing systems.

One way out of this dilemma is to significantly increase the sensitivity of the receiver once again such that less laser intensity is needed. Another way out of this dilemma is to switch from 905 nm NIR technology to 1550 nm SWIR technology. On the one hand, tests (Kutila et al. 2016; Mäyrä et al. 2017) have shown that the penetration of fog in the SWIR band is better than in the NIR band. More importantly,

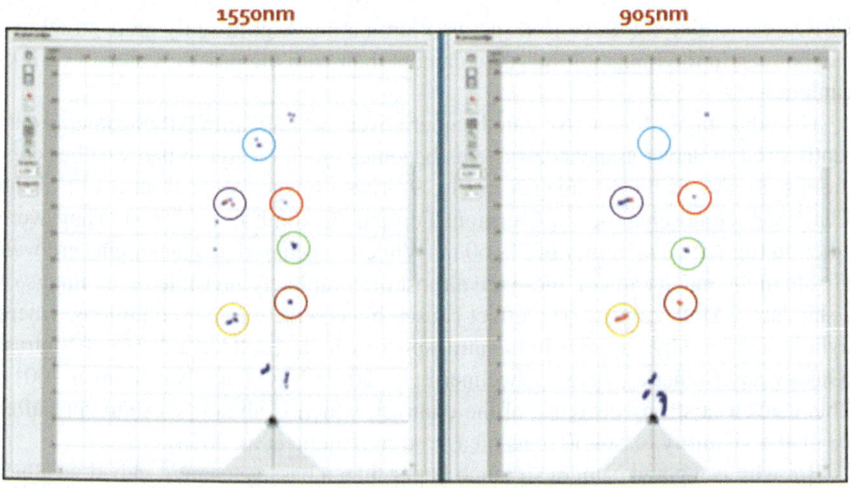

Fig. 7 Fog chamber test result: Fog penetration @ Visibility range 55 m. Echoes from reflectors and persons walking around. 1st echos = blue, 2nd echos = red. The 1550 nm Lidar clearly outperforms the 905 nm Lidar of identical construction. While in the 905 nm Lidar most first echoes are reflected from the fog wall and the objects are usually only detected with the second echo, the 1550 nm Lidar penetrates the fog wall and detects the objects with the first echo

the laser in the SWIR band can have up to 40 times more energy than in the NIR band and is still eye safe.

For this reason, we are building a 1550 nm Lidar in the DENSE project. Initial preliminary studies have shown that the massive amount of optical output power, that is allowed to be emitted at 1550 nm due to relaxed eye safety regulations at higher wavelengths clearly increases the Signal-to-Noise Ratio (SNR) of the Lidar's time-of-flight signal significantly (Holzhüter et al. 2018).

Tests in the parallel project RobustSENSE (see http://www.robustsense.eu/) on the basis of a single-beam Lidar from Sick have shown that a 1550 nm Lidar clearly outperforms the identical 905 nm Lidar, which could be demonstrated on the final event of the RobustSENSE project (Dietmayer and Waldschmidt 2018). An example of a result from a fog chamber test with these Lidars is shown in Fig. 7. Similar results are expected in DENSE with the completion of the project. In contrast to the RobustSENSE Lidar with only one laser beam, a multi beam Lidar will be developed and shown.

4 Gated Camera

Instead of a passive CMOS camera we want to use an active gated camera for our DENSE platform. As you can see in Fig. 8, the prototypically realized gated camera from BrightWayVision (Inbar and Ofer 2008) already has advantages over a passive

CMOS camera at night due to its active illumination. Significantly more structures can be seen in the image of the gated camera than with the passive standard CMOS camera.

The advantage of an active gated camera over the standard CMOS cameras currently used in driver assistance systems becomes much clearer in bad weather conditions, as shown by the images in Fig. 9. These images were from a test in the CEREMA cloud chamber in Clermont-Ferrand (Colomb et al. 2008). They were taken in fog with a visibility of 50–60 m. While the pedestrian mannequins are well visible in the gated camera, these are almost or completely invisible in the image of a standard CMOS camera. The effect caused by adverse weather conditions, where light is scattered by particles in the atmosphere which results in a decay of contrast (Oakley and Satherley 1998; Schechner et al. 2001; Nayar and Narasimhan 2001). This leads to a strong decrease of the contrast, which is already close to zero after 20 m at a visibility of 50–60 m as measured in (Bijelic et al. 2018).

But why is a gated camera so much better than a passive camera in bad weather. This is where the gating mechanism comes in. The trick with a gated camera is to illuminate and record only a defined depth range. Disturbing reflections from e.g. fog, which occur outside of this depth range are not recorded because the gate is closed and thus suppressed. In contrast to this, a passive non-gating camera records

(a) Standard CMOS camera *(b) Gated camera*

Fig. 8 Passive standard CMOS color camera versus active gated camera at night: Due to the active illumination, significantly more structures can be seen in the image of the gated camera than with the passive standard CMOS camera

(a) Standard CMOS camera *(b) Gated camera*

Fig. 9 Raw images of a standard CMOS camera and a gated camera taken in artificial fog in a fog chamber

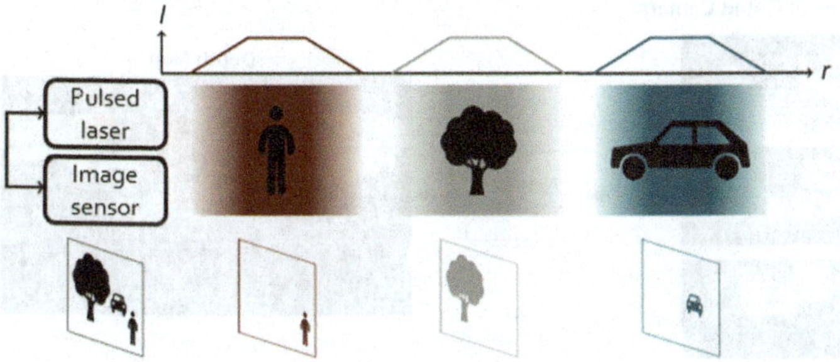

Fig. 10 The gating mechanism allows depth areas to be explicitly illuminated and recorded. Disturbing reflections outside the depth range, such as fog, are not included and thus suppressed. The overall image is formed from the individual depth ranges by superimposition

Fig. 11 Range intensity profiles (RIPs) of our gated camera used to generate the depth map of Fig. 12

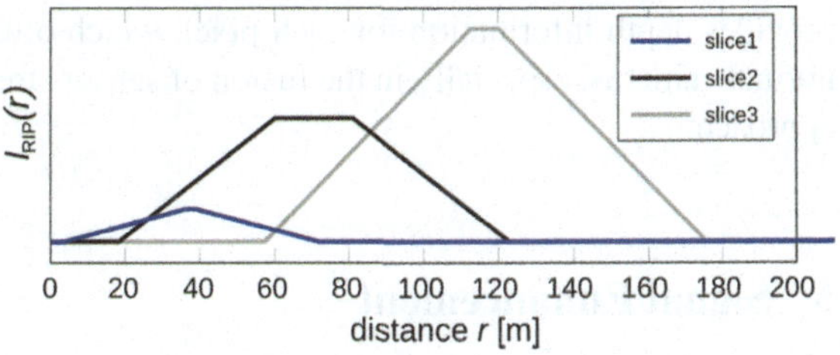

the disturbing reflections, which in turn superimpose important image contents and thus become invisible.

The gating mechanism is more detailed explained in (Gruber et al. 2018). In general, our gated camera records several depth ranges one after the other and then generates a result image from these range images by superimposing these images. In Fig. 10 this is shown as an example for 3 depth ranges. Since the exposure times are in the ns range, our gated camera can record up to 4 depth ranges with a resulting refresh rate of 30 Hz.

The upper part of Fig. 10 shows the illumination profiles over the distance, the so-called Range-Intensity-Profiles (RIP), in the form of spatially separated trapezoids. In reality, we do not use separate depth ranges but overlay them using different RIPs, as shown in Fig. 11.

This allows us to create a depth map from the depth images. We use a neural network to calculate this depth map. Detailed information on the calculation of the depth map can be found in (Gruber et al. 2018). Figure 12 shows a result image of this depth map. The advantage of this approach is, that the range estimation is independent from the Range Intensity Profile (RIP) and the target reflectivity and that only three overlapping depth range images are required for a relative accuracy of 5%

With the gated camera we are taking a significant step towards an all-weather 24/7 perception system. In addition to the image information, the gated camera also

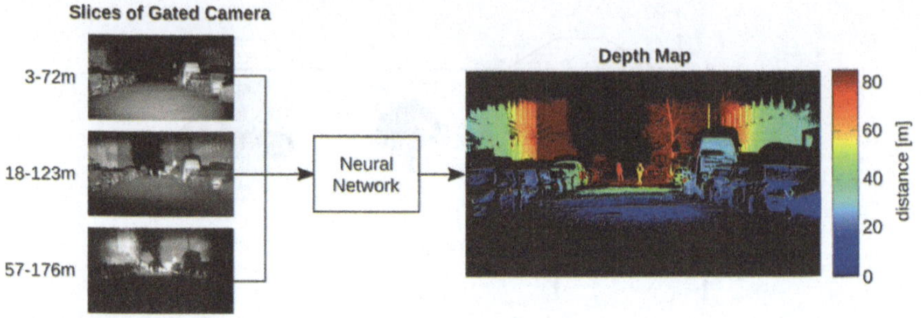

Fig. 12 Pixel-accurate depth map generated via a neural network from the depth images of the gated camera

provides depth information for each pixel, which offers enormous advantages over normal cameras, especially in the fusion of sensor streams envisaged in the DENSE approach.

5 Signal Enhancement

The weather-related deterioration of the sensor signals generally leads to a reduction in the perception performance of an ADAS system. This reduction in perception can either be compensated by improved sensor technology, as described in the previous sections, or through intelligent post-processing.

In this chapter we present two examples of intelligent post-processing: Use of an additional sensor stream to compensate for the disturbance, and the restoration of disturbed information from a disturbed sensor.

5.1 Use of An Additional Sensor Stream

One often used method to increase the detection performance is to fuse the raw data signal with neural networks that learn robust features from multiple inputs. In case of a disturbed sensor stream the fusion is not so easy anymore (or straight forward to do), because the variations and complexity of the disturbances are so great that they can never be fully registered and learned. Neural networks performing a Feature Level Sensor Fusion can be particularly error prone to disturbances outside the training data distribution, even if the other undisturbed sensor stream provide high-quality data (Bijelic and Münch 2018).

To overcome this problem, we propose a simple data augmentation scheme that ignores data from degraded sensors, although it has never seen these degraded characteristics of this sensor during training. This can be summarized as a learned "OR"

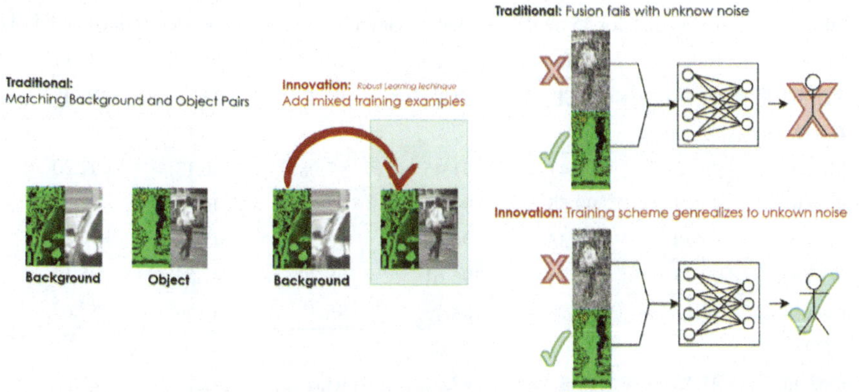

Fig. 13 Proposed learning technique: Training a fusion network with mixed examples out of background and object class enables the network to learn a 'or' operation between the two input channels. This enables to retrieve information from one channel even if the second channel is disturbed with unknown disturbances

operation at the fusion stage. This learned operation generally applies to all types of disorders that do not occur during training.

The learning scheme for increasing robustness is as follows (see also Bijelic and Münch 2018): We first train a dual-stream basic fusion network with a sufficiently large set of undisturbed data pairs. In order to make it insensitive to disturbances, we now train in a second step the network with data pairs where one element (stream) contains data of the background class and the other element (stream) contains data of the undisturbed object class. Already with a share of 10–20% of disturbed data pairs we generate a fusion network, which is robust against disturbances in a stream. In our experiments, the percentage of disturbed data pairs in the base network could be up to 50% without significantly changing the behavior of the fusion network (Fig. 13).

We compared the traditional learning technique (TRL) with our new learning scheme, the robust learning technique (RLT) using the KITTI [KITTI] dataset and the Gavrilla dataset (Enzweiler et al. 2010). For this purpose, we degraded these data sets using Gaussian Noise RGPN ($\sigma = 0.9$) and Gaussian Blur GBLUR ($\sigma = 0.9$). As can be seen from Table 2, the performance capability of the network trained with robust learning technique (RLT) hardly differs from that of the network trained with the traditional learning technique (TLT) with undisturbed samples (Ref.). However, the difference between these networks in the classification of disturbed data sets becomes significant. Here, the network trained with the robust learning technique (RLT) clearly performs better than the network trained with the traditional learning technique. A detailed description of this new learning technique can be found in (Bijelic and Münch 2018).

Table 2 Recall Statistics for 45 models trained on Gavrilla-Data and 63 models trained on KITTI

Noise	Ref.	Ref.	GBLUR	GBLUR	RGPN	RGPN
RLT/TLT	TLT	**RLT**	TLT	**RLT**	TLT	**RLT**
GavData						
Median	99.55	99.41	91.45	97.08	46.13	92.20
u-quart.	99.64	99.55	92.44	97.88	46.61	93.50
l-quart.	99.48	99.33	90.69	96.36	45.64	90.86
u-whisk.	99.73	99.73	96.61	98.70	47.54	95.10
l-whisk.	99.17	98.78	88.73	95.46	40.54	87.63
KITTI						
median	97.1	97.3	83.5	93.0	42.7	86.2
u-quart.	97.3	97.5	84.7	94.4	48.6	89.6
l-quart.	96.9	97.2	81.1	91.1	30.3	78.4
u-whisk.	97.7	97.7	86.8	95.8	54.8	95.4
l-whisk.	95.0	96.6	72.7	89.7	8.5	68.4

Hard numbers from RLT and TLT corresponding to Fig. 5. The reference (Ref.) shows the recall for undisturbed data

5.2 *Restoration of Disturbed Sensor Information*

Another possibility to increase the performance of the perception system under bad weather conditions is to restore the sensor data affected by fog, snow and rain, i.e. to restore the disturbed sensor stream.

The problem here is that currently all available data sets for automated driving such as KITTI and Cityscapes are based on good weather conditions. The generation of a bad weather data set in the same way as for the generation of KITTI and Cityscapes seems to be complex and expensive, because bad weather conditions generally cannot be caused by the push of a button, and certainly not with defined bad weather characteristics, e.g. a defined visibility or a defined rain intensity, so that the acquisition of a complete (i.e. sufficiently large) data set appears very difficult to impossible. For this reason, in order to generate a learning data set for automated driving in bad weather, a data extension scheme should be used which converts good weather data into bad weather data.

In our case we relayed on simulated foggy driving scenes based on the KITTI and Cityscapes dataset, which can be easily extended to weather such as snow and rain.

This is one of the key ingredients for the following neural networks, since learning methods usually use a learning pair (undisturbed example, disturbed example) for training, which can also be generated much more easily in simulation than in real driving scenes, where for example the same image has to be captured once undisturbed and once e.g. in fog (with a defined visibility).

Fig. 14 Original image from the Cityscape data set and the result images of the fogging algorithms simulated for visibility ranges of 800, 400, 200 m

The fogging algorithm is based on (Sakaridis et al. 2018), which we have modified for our purposes. Result images of the fogging algorithm can be found in Fig. 14.

For de-fogging we used a global adversarial network (GAN). We rely on the GAN developed by (Wang et al. 2018), called Pix2PixHD, which we have modified and extended according to our needs (e.g. by adding an inversion architecture, a Residual Architecture, a Gradient Clipping, etc.). Further information can be found in (Bijelic et al. 2019).

The defogging algorithm developed in (Bijelic et al. 2019) was applied to artificially fogged images and controlled fogged scenarios in a fog chamber and real fogged scenes in cities. The results were very promising, covering a different range of different evaluation matrices as SSIM (Wang et al. 2004), LPIPS (Zang et al. 2018), VIF (Sheik and Bovic 2005), etc. A qualitative enhancement result can be seen in Fig. 15.

Furthermore the defogging algorithm is able to significantly enhance detection results of downstream "good-weather" detection networks. To show this, we applied a Single Shot Multi-Box Detector (Liu et al. 2015) for vehicle detection on the fogged KITTI data set with and without our defogging network as defogging preprocessor. The promising results of this experiments are listed in Table 3. The application of the de-fogging network as preprocessor led to significantly better detection results for fogged images.

Fig. 15 Results of the defogging algorithm applied on real fogged images (not belonging to the learn set)

6 Conclusions

This paper has presented the system architecture and development results of three key components of the DENSE sensor suite. These were the Lidar, the gated camera and two signal enhancement methods for the processing unit realized by neural networks.

In tests with the Lidar it could be shown that existing 905 nm Lidars degrade strongly in bad weather. In this case the change from 905 nm to 1550 nm offers a remedy. First tests have shown that 1550 nm Lidar significantly outperformed the 905 nm Lidar.

As with the 905 nm Lidar, the CMOS camera currently used as standard in driver assistance systems degrades strongly in bad weather. The development of the gated camera presented in this paper, on the other hand, proved to be robust against bad weather and also delivers a pixel-accurate depth map which offers enormous advantages over normal cameras, especially in the fusion of sensor streams envisaged in the DENSE approach.

Finally, two signal enhancement methods were presented for the processing unit of the DENSE Sensor Suite, which is realized by deep neural networks. On the one hand, this was new (learning) procedure with which multi sensor fusions networks can be made robust against disturbances of a disturbed sensor signal and, on the other hand, an adversarial neural network for defogging of fogged input images. The defogging approach shown here can be extended to restore sensor signals from other types of disturbances such as caused by snow and dense rain.

The components described in this paper mark an important step towards a 24/7 all-weather perception system for driver assistance systems. The DENSE sensor suite and its components will be further developed until the end of the project in spring 2020.

Table 3 A Single Shot MultiBox Detector (SSD) (Liu et al. 2015) was applied to foggy images and defogged images using our defogging network

Model	Fog density								
	β	0	0.005	0.01	0.02	0.04	0.06	0.08	0.16
	MOR	No fog	600 m	300 m	150 m	75 m	56 m	37 m	19 m
Without preprocessor	Easy moderate hard	74.64	75.02	75.80	77.60	80.56	83.34	75.16	22.55
		81.64	82.14	82.55	82.44	75.88	08.87	58.59	15.99
		65.36	65.86	66.20	66.39	66.63	59.54	50.21	15.77
Our model M_{cj}	Easy moderate hand	76.46	76.55	76.72	77.08	77.63	78.38	79.77	**74.79**
		82.02	82.03	82.10	82.07	81.48	75.36	75.57	**50.91**
		65.89	65.83	65.87	65.93	66.20	66.43	66.12	**42.59**

Average Precision (AP) scores measured for three variants. For the variant "without preprocessor", the detector is directly applied to all foggy images. The same detector scores are calculated for the output images O_{cj} of the defogger M_{cj}. The AP scores are specified for different level of fog and detectability of objects. The detector model provides bounding boxes of detected cars. For each bounding, box the model computes a probability of how likely a car is inside the bounding box. If the probability is greater than a threshold, the model outputs the bounding box as a detected car. The threshold is parameterized. Different thresholds settings result in different Precision and Recall scores. The AP metric is an estimator for the area below the curve between 0 and 1 Recall. Therefore, the AP metric is an expected value of Precision dependent on Recall

Acknowledgements We would like to express our gratitude to the whole DENSE project (Grant Agreement number: 692449) Consortium for fruitful discussions on the designing of the DENSE architecture and its components. The project has been co-funded by ECSEL Joint Undertaking and national funding agencies like Business Finland and Federal Ministry of Education and Research in Germany.

References

Bijelic M, Münch C et al (2018) Robustness against unknown noise for raw data fusing neural networks. ITSC

Bijelic M, Gruber T, Ritter W (2018) A Benchmark for lidar sensors in fog: is detection breaking down? In: IEEE intelligent vehicles symposium

Bijelic M, Gruber T, Ritter W (2018) Benchmarking image sensors under adverse weather conditions for autonomous driving. In IEEE Intelligent Vehicles Symposium

Bijelic M, Mannan F, Gruber T, Ritter W, Dietmayer K, Heide F (2019) Seeing through fog without seeing fog: deep sensor fusion in absence of labeled training data. In: CVPR 2019, under review. Longbeach, CA, USA

Colomb M, Hirech K, Andr P, Boreux J, Lacte C, Dufour J (2008) An innovative artificial fog production device improved in the European project fog. Atmos Res 87(34):242–251

Dietmayer K, Waldschmidt C et al (2018) "http://www.robustsense.eu/," RobustSENSE final event Ulm, final event presentations, 04_RobustSENSE_WP3_ 20180516_Dietmayer_Sensor and fusion layer.pdf, 16 May 2018. Available: http://www.robustsense.eu/. Accessed 16 Jan 2019

Enzweiler M, Eigenstetter A, Schiele B, Gavrila D (2010) Multi-cue pedestrian classification with partial occlusion handling. CVPR, San Francisco

Gruber T, Kokhova M, Ritter W, Haala N, Dietmayer K (2018) Learning super-resolved depth from active gated imaging. ITSC

Holzhüter H, Kutila M, Pyykönen P, Colomb M (2018) Automotive LiDAR performance verification in fog and rain. In: IEEE international conference on intelligent transportation systems (ITSC), Maui, USA

Inbar S, Ofer D (2008) Laser gated camera imaging system and method. US Patent 7,379,164

Källhammer J, Cassel A (2017) Overall system requirements. EU DENSE ECSEL project Deliverable 2.3

Kutila M, Pyykönen P, Ritter W, Sawade O, Schäufele B (2016) Automotive LIDAR sensor development scenarios for harsh weather conditions. In: IEEE 19th international conference on intelligent transportation systems (ITSC), Rio de Janeiro, Brasil

Liu W, Anguelov D, Erhan D, Szegedy C, Reed S, Fu C, Berg C (2015) SSD: single shot multibox detector. In: CoRR, vol. abs/1512.02325. Available at: http://arxiv.org/abs/1512.02325

Mäyrä A, Hietala E, Kutila M, Pyykönen P, Mäyrä A, Hietala E, Kutila M, Pyykönen P (2017) Spectral attenuation in low visibility artificial fog: experimental study and comparison to literature models. In: 13th IEEE international conference on intelligent computer communication and processing, Cluj-Napoca, Romania

Nayar S, Narasimhan S (2001) Removing weather effects from monochrome images. Conf Comput Vision Pattern Recog 2:186–193

Oakley JP, Satherley BL (1998) Improving image quality in poor visibility conditions using a physical model for contrast degradation. IEEE Trans Image Process 7(2):167–179

Sakaridis C, Dai D, Van Gool L (2018) Semantic foggy scene understanding with synthetic data. Int J Comput Vision

Schechner Y, Narasimhan S, Nayer S (2001) Instant dehazing of images using polarization. In: Conference on computer vision and pattern recognition, vol 1, pp 325–332

Sheik H, Bovic A (2005) A visual information fidelity approach to video quality assessment. In: The First International Workshop on Video Processing and Quality Metrics for Consumer Electronics

Wang Z, Bovic C, Sheikh H, Simoncelli E (2004) Image quality assessment: from error visibility to structural similarity. In: IEEE transactions on image processing

Wang T-C, Liu M-Y, Zhu J-Y, Tao A, Kautz J, Catanzaro B (2018) High resolution image synthesis and semantic manipulation with conditional gans. In Proceedings of the IEEE conference on computer vision and pattern recognition

Zang R, Isola P, Efros A, Shechtman E, Wang O (2018) The unreasonable effectiveness of deep features as a perceptual metric. In IEEE Conference on Computer Vision and Pattern Recognition (CVPR)

[illegible] (2017) [illegible] image-to-image translation using cycle-consistent adversarial networks. In: IEEE International Conference on Computer Vision (ICCV)

Are LIDARs Ready for Perception in Future Intelligent Transportation?

Diego Puschini, Cem Karaoguz, Oussama El-Hamzaoui and Tiana Rakotovao

Abstract The LIDAR is seen as a key enabling technology for future autonomous transportation. The specification and performance of LIDAR devices are crucial to enable safe perception algorithms using a combination of several sensing technologies. Nevertheless, do existing LIDAR devices fulfill automotive expectations? We have tested a set of LIDARs covering the most representative technologies available on the market. This text reviews some relevant results showing limitations and advantages of this technology in the context of autonomous vehicles.

Keywords LIDAR · Perception · Data fusion · Autonomous vehicles

1 Introduction

Nowadays, LIDAR technologies are evolving, mainly driven by the intelligent transportation industry. During year 2017, more than 350 million dollars have been invested on a few small and medium companies developing LIDAR sensors (LiDARs for Automotive and Industrial Applications 2018). This amount of investments have been increased in 2018, showing clearly that available LIDAR technologies are still evolving. The reason is that LIDAR is seen as a key enabling technology for future autonomous transportation. These sensors combined with existing perception technologies like radar and vision, are the eyes of future autonomous vehicles (Rakotovao et al. 2016) (Broggi 2013; Aeberhard 2015). The specification and performance of these devices are decisive to enable safe autonomous driving through perception algorithms using a combination of these technologies.

Are existing LIDAR devices fulfilling automotive expectations? A set of LIDARs covering some of the most representative technologies available on the market have

D. Puschini (✉) · T. Rakotovao
CEA-Leti, Technology Research Institute, Minatec Campus, Grenoble, France
e-mail: diego.puschini@cea.fr

C. Karaoguz · O. El-Hamzaoui
Transdev, Issy-les-Moulineaux, France

J. Langheim (ed.), *Electronics Components and Systems for Automotive Applications*, Lecture Notes in Mobility,
https://doi.org/10.1007/978-3-030-14156-1_13

been tested. The objective of these tests is to firstly understand LIDAR advantages and limitations, second to enable choosing the right set of LIDARs to be used on novel intelligent vehicles and finally to provide feedback to sensor technology developers in order to increase sensor performance and coverage. The tested sensors includes solid-state flash, mechanic scanning and hybrid flash/mechanic scanning LIDARs. Some of the tested technologies are widely used in prototypes of autonomous vehicles all around the world and are seen as serious candidate to be widely deployed.

Sensors were tested under the same real world conditions through specific methodologies using generic metrics. Several targets were explored under different weather conditions (sun, rain, snow, fog). The tested targets were selected in order to cover a set of typical obstacles typically found by autonomous vehicles on real situations.

This study has been carried by a collaboration between Transdev,[1] a leading global provider of mobility service and an active actor of intelligent transportation, and CEA-Leti,[2] a leading French research institute, providing innovation in micro and Nano electronics, MEMs and sensors.

2 LIDAR Sensors for Intelligent Transportation

An autonomous vehicle can be represented by three main functions: *sense*, *think* and *act*. These three functions are critical to guarantee the whole system behavior: they define the interaction between the autonomous vehicle and the environment. In this scheme, specific algorithms build virtual models of the environment where the required information is represented based on the information provided by sensors in order to control the robot. Sensors are in charge of feeding the information to build virtual environment model, playing a critical role in the whole system.

Autonomous vehicles require 360° perception. There exist several candidate sensors to consider as a complement to provide full perception. The main families are image sensing (visible and/or infrared), ultrasound, radar and LIDAR (Van Brummelen et al. 2018). Each technology offers a set of advantages and disadvantages that need to be taken into account to ensure safe perception. LIDAR are seen as offering high-resolution, dense and precise range measurements. It is currently used as the main key technology to enable autonomous driving. However, in order to analyze their current maturity some questions need to be addressed in the context of autonomous intelligent transportation.

[1]https://www.transdev.com.

[2]http://www.leti-cea.fr.

2.1 Covered Distance

At the present time, several LIDAR companies claim detection ranges up to more than 100 m. Nevertheless, the maximal detection distance strongly depends on several factors such as the lighting source, the photonic detector, the scanning mode and the optical system of the sensor, as well as the reflectivity of the detected object. What is the real perception distance of the current generation of LIDARs?

2.2 Sensor Cross-Talk

As explained above, a vehicle represented by the scheme *sense-think-act* becomes a more complex system when the external environment is taken into account. Considering that the road will be shared with other users, including other autonomous vehicles equipped with LIDARs, a cross-talk problem could potentially appears, strongly impacting the perception performance. What happens if several LIDARs share the roads? Are there any conflictive situation where LIDAR performance is affected by other sensors or perturbation sources in the environment?

2.3 Impact of Weather Conditions

Vehicles are supposed to run under all weather conditions. Water drops, snowflakes or particles in the air due to dense fog can potentially impact the detection performance of LIDARs. All these elements are seen as perturbations and should be considered in the perception problem of vehicles equipped with LIDARs. Are current LIDARs robust enough under all weather conditions? What are the real performance of sensors under strong rain, snow or dense fog weather?

3 Tests and Methodology

LIDAR devices produce measurements that are related to the distance to the sensed targets. However, LIDAR measurements are noisy. Measurement noises can be caused by various origins such as imperfection of electric/electronic components within the sensor, numerical errors during digital signal processing, external lighting conditions, weather conditions, etc.

The objective of the present work is to analyze the effect of the distance to the target, the presence of cross-talk and the weather conditions on LIDAR measurements. For this purpose, experiments were realized under a rigorous methodology. The methodology enables to test multiple LIDAR devices under the same experimental conditions, to compare the results between multiple devices and to make experiments reproducible for both existing and future devices.

3.1 Experimental Methodology

The experimental methodology consists in putting a target at a known distance from a sensor under test. Such test is repeated under various experimental conditions. A population of measurements from the sensors are then recorded during a measurement campaign and processed further in order to extract statistical metrics that enable to analyze the effect of each experimental condition on the measurement.

3.2 Experimental Conditions

During an experiment, a target is placed in the front of the sensor under test. The tested targets are presented in more detailed in the following section. A laser telemeter with high-accuracy measures the distance from the sensor under test to the target. The telemeter acts as a reference instrument for measuring distance and it has been calibrated by an external independent laboratory.

Experiments were realized under four weather conditions: sunny, rainy, foggy and snowy. The objective is to verify the impact of rain, fog and snow on LIDAR measurements and to compare the performance of different LIDAR devices under various weather conditions.

To analyze the effect of cross-talk, another LIDAR device is placed in the front of the sensor under test together with a target. The other LIDAR acts as an assailant, source of potential interference. When turned on, the assailant emits laser beam that can disturb the process of measurement on the sensor under test.

3.3 Targets

To measure range, a LIDAR emits a laser beam toward a target. After hitting the target, the beam travels back to the LIDAR. Its time of flight enables then to estimate the range of the target. It is crucial for the LIDAR to be able to detect properly the returning beam in order to measure the time of flight. The reflectivity of the target affects the quantity of light echoed back to the LIDAR. Therefore, reflectivity also has an effect on the accuracy of LIDAR measures.

While a LIDAR mounted on a car may encounter targets of diverse natures, the experiments were realized on four targets ranging from low to high reflectivity. The tested targets listed in increasing reflectivity are a tire, a cardboard, a portion of the

Fig. 1 Tested targets from left to right: a tire (low reflectivity), a cardboard (medium reflectivity), a portion of a car body (high reflectivity) and reflective tapes (light reflector)

body of car and a reflector surface similar to road signals. These targets are also representative of objects commonly found on real road situations.

As presented on Fig. 1, the tire has a low reflectivity due to its color, its manufacturing material, the ruggedness of its surface. The cardboard has a higher reflectivity but still lower than the reflectivity of the car body. The road signal acts as a light reflector. It reflects a light beam within the same direction of the incoming beam, whatever is the incidence angle.

3.4 Experimental Setup and Data Campaign

A set of LIDAR devices already existing on the market are selected for the study. Experiments are realized under the combinations of the following conditions: nature of target, distance to target, presence of cross-talk, and weather conditions. For each combination, a population of measurements (typically 5000 samples) from each LIDAR device are recorded on a computer and processed offline for further analysis.

The reference telemeter and the LIDARs to be tested are mounted on a support composed of a fixed lower part and a mobile upper part. The fixed part is a tripod which role is to support the upper part and to maintain it at a known height from the ground. The upper part is a horizontal bar that can be slid to the left or to the right. The LIDARs and the telemeter are screwed on the upper part.

The experimental setup is illustrated on Fig. 2. The sensor under test and the target are positioned as follows. First, the target is put in the front of the reference telemeter at a known distance. This distance represents the ground truth. Second, a LIDAR is selected as the sensor under test by sliding the upper part of the support. The support ensures that the distance between the sensor under test and the target remain unchanged. After that, a set of measurements from the sensor under test are recorded on a computer. Then, the next LIDAR is tested by sliding the upper part of the support and repeat the same steps as for the last sensor under test.

The above approach allows to ensure three points. First, the ground truth is known and can be adjusted in advance. Second, the sensors are tested under the same conditions, that is, the same target, distance, presence of cross-talk, and weather condition. Third, the experiments are reproducible.

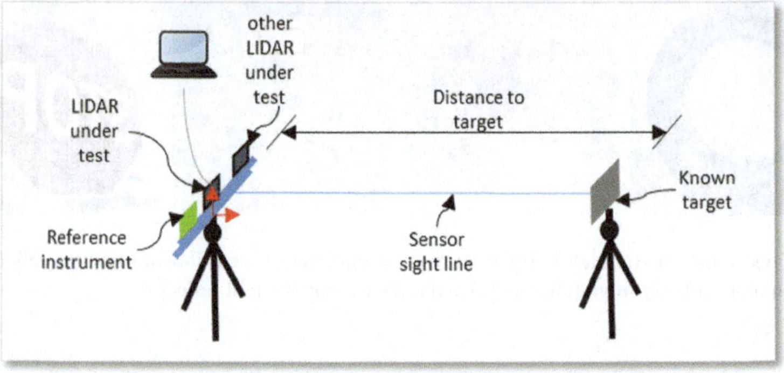

Fig. 2 Experimental setup

4 Result Examples

A set of sensors available on the market have been chosen in order to cover different LIDAR architectures. These sensors have been tested following the methodology detailed in the previous section. The chosen sensors are representative of the current LIDAR scanning architectures available on the market: (A) hybrid flash and scanning, (B) 360° scanning, (C) full flash and (D) scanning less than 360°. Without unveiling LIDAR trades, the next sections describe some observed issues and limitations under the different evaluated conditions.

4.1 Nominal Conditions

The first set of experiments were performed at nominal conditions, under sunny weather and with no external cross-talk. Under these conditions, one can suppose that LIDARs should obtain good to excellent performances.

Figures 3 and 4 are examples of the detection rate of some evaluated LIDARs with two different targets, the vehicle tire and the portion of a car body, at different distances up to 100 m. Figure 3 clearly shows that tested LIDARs were unable to detect the low-reflective target at distances higher than 35 m. The supposed reason for these very low performance results is that the amount of light returned by the target is not enough to be detected at higher distances.

Figure 4 shows the detection rate for a portion of a vehicle as a target. Some LIDARs were unable to detect the target for distances higher than 50 m, showing a real limitation. Only one based on partial scanning architecture (D) was able to perfectly cover up to 100 m with the car body target.

For this target, one can suppose that the amount of energy returned by the target should be enough for higher distances. However, results show another reality. For mechanical scanning devices, the angular dispersion of the couple source-detector

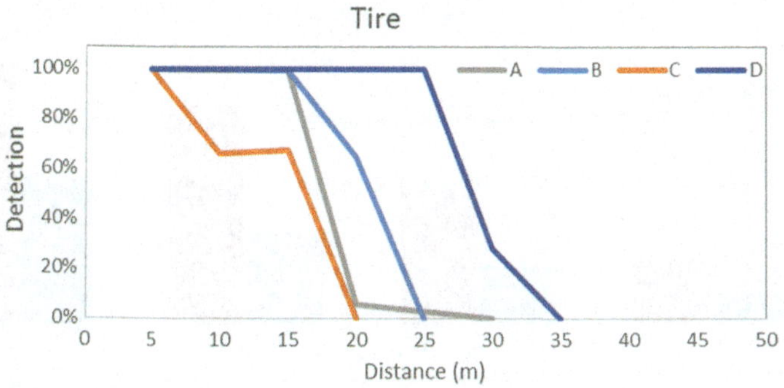

Fig. 3 Example of detection of a tire at different distances up to 100 m

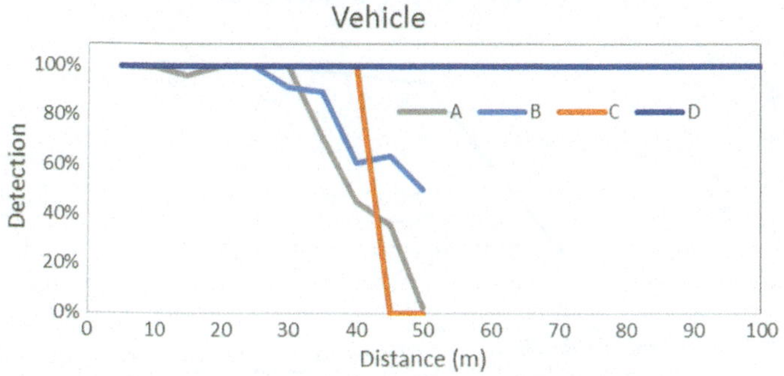

Fig. 4 Example of a vehicle detection at different distances up to 100 m

increases the number of missing or false detections. In other words, the detection rate strongly decrease with the target size. Systems with partial scanning are potentially offering better angular dispersion, reducing the number of missing and/or false detections. In the case of flash LIDARs, the amount of maximal emitted energy is supposed to be spread to cover a given sector. Thus, even if there is no angular dispersion, the amount of energy being returned to the light detector becomes the main limitation on the detection distance.

Figure 5 shows a comparison between the specifications and the obtained results for the two different targets. Each bar of this figure corresponds to the ratio between the distance obtaining 50% of detection rate over the maximal distance specified by the LIDAR provider. The gaps between the obtained results and the specifications given by LIDAR manufacturers can be justified by the difference between the tests conditions and methodology used in this study, compared to the ideal and optimist scenarios evaluated by sensor providers. Nonetheless, if these devices are used for real applications, performance should be closer to the ones obtained on this study, being more realistic than ideal cases.

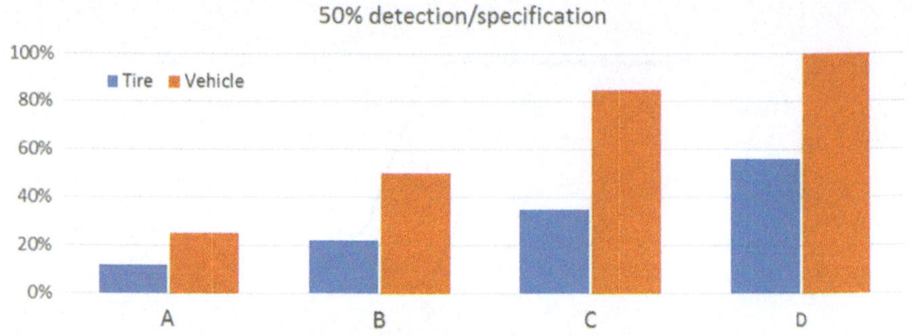

Fig. 5 Detection distance compared to specifications

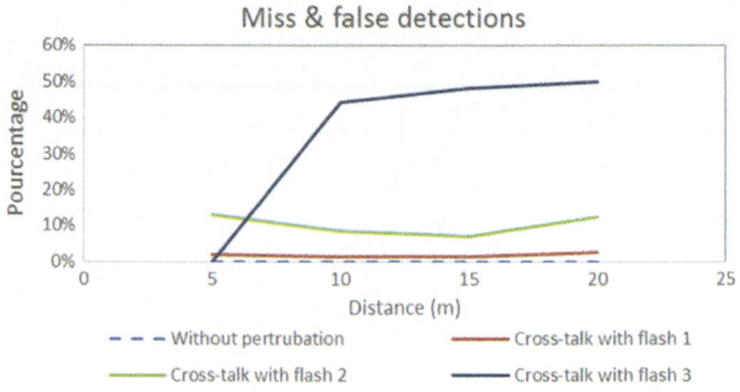

Fig. 6 Increasing miss and false detections due to cross-talk

4.2 *Cross-Talk Analysis*

The cross-talk analysis clearly revealed three identified impacts on the performance of sensors. Firstly, a strong increasing on the miss and false detection have been observed. Miss detections correspond to scenarios where the sensor is unable to perform a detection, the sensor has not been able to identify a clear light echo coming from the environment. On the contrary, false detections correspond to aberrant measurements, for instance when the output is 2 times or more the real distance to the target. Figure 6 shows an example of this behavior for a scanning sensor.

The second observed impact concerns the sensor precision under cross-talk conditions. The precision has been evaluated through the standard deviation of a population of measurements performed under the same conditions. An increase (sometimes strong) of the standard deviation has been observed, indicating a decrease of the sensor precision in term of distance. Figure 7 presents an example of this effect for a scanning LIDAR under cross-talk situations with three different flash LIDARs, evaluated at different distances between 5 and 20 m.

Finally, the last observed behavior is the decreasing accuracy of distance measurements. The accuracy has been evaluated through the error of the mean value of

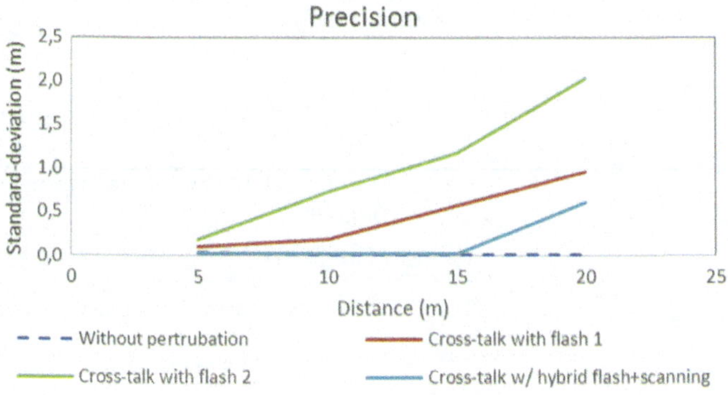

Fig. 7 Cross-talk impact on sensor precision

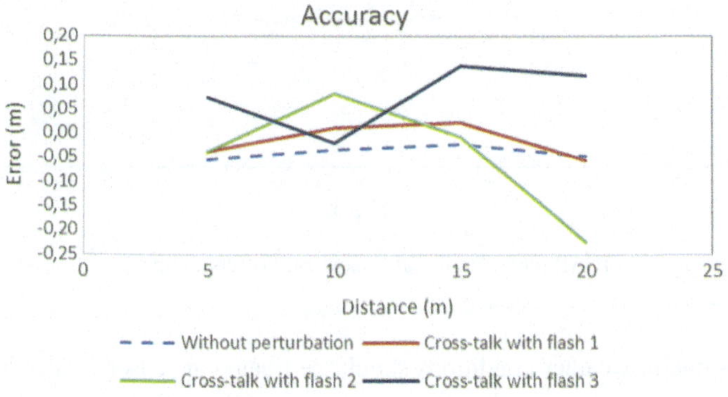

Fig. 8 Cross-talk impact on sensor accuracy

a representative population of measurements performed under the same conditions. Figure 8 shows an example of a sensor where the error without a cross-talk is less than 5 cm. However, the error can reach up to 25 cm under cross-talk situations.

4.3 *Impact of Weather Conditions*

Some tests were repeated under different weather circumstances, including strong rain, snow and dense fog. Compared to nominal conditions, the main impact is visible on an important increasing of miss and false detection rate. Figure 9 shows detection histogram for a scanning LIDAR with a tire target, typically perfectly detected at 20 m in clear weather. Under rain condition, the detection rate is lower than 20% due to miss detections. Under snow, the detection rate is below 10% with several false detection between 0.5 and 6 m produced by the snowflakes. Under fog conditions, no more detection is possible for this set-up. This example clearly shows that false

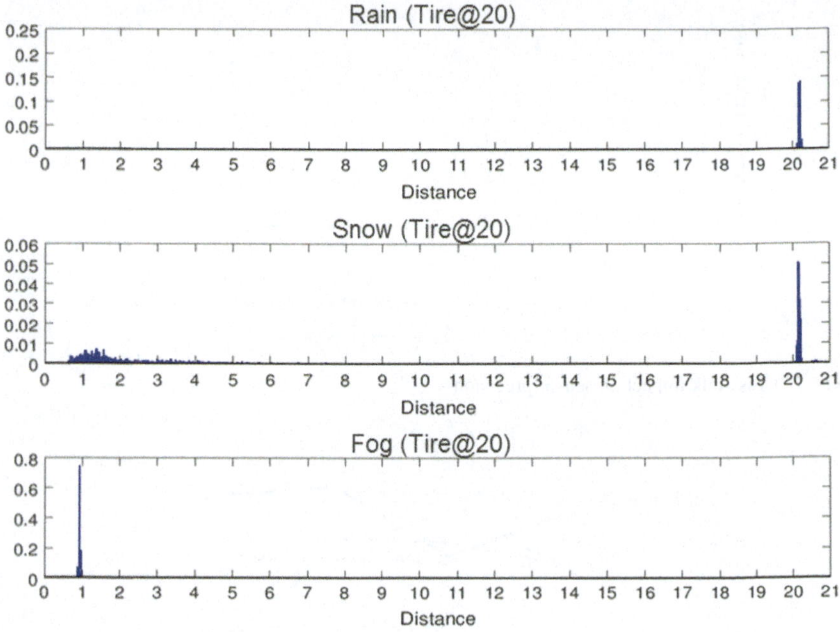

Fig. 9 Example of a LIDAR under strong rain, snow and fog conditions

detections due to weather conditions should be treated at sensor or algorithm level to improve the perception performance.

5 Conclusion

LIDARs are seen as key enabling technology for high level of autonomy in automotive and intelligent transportation. They are often responsible of an important part of the environment sensing, and thus, of the vehicle and users safety since critical actions are performed taking into account the information they can provide. The experiments performed during this work put in evidence some important lacks on current LIDAR sensors in terms of detection distance under real conditions. The detection performance strongly varies depending on the test conditions and, most of the times, they are far from ideal performance proposed by sensor providers.

Today, there is no standardized tests to characterize sensors dedicated to the environment perception for automotive applications. Every company can unilaterally claim anything in terms of detection performance without any kind of standardized procedure. For sure, this is something that should evolve. What are the representative criteria that should be evaluated? What are the test procedure and conditions that perfectly cover all the representative criteria? These questions should be addressed by all the actors in the automotive value-chain.

This study clearly highlighted the necessity to continue the roadmap of LIDAR sensors. More and more, several companies are proposing approaches where the performance problem is tacked through embedded algorithms, coupling it with other sensing technologies and turning the sensor into a smart one. This is a potential and credible way to keep the LIDAR as key enabling technology in autonomous transportation.

References

Aeberhard M et al (2015) Experience, results and lessons learned from automated driving on Germany's highways. IEEE Intell Transp Syst Mag 7(1):42–57. https://doi.org/10.1109/mits.2014.2360306

Broggi A et al (2013) Extensive tests of autonomous driving technologies. In IEEE Transactions on Intelligent Transportation Systems, vol 14, no 3, pp 1403–1415. https://doi.org/10.1109/tits.2013.2262331

LiDARs for Automotive and Industrial Applications 2018. http://www.yole.fr

Rakotovao T, Mottin J, Puschini D, Laugier C (2016) Multi-sensor fusion of occupancy grids based on integer arithmetic. In: 2016 IEEE international conference on robotics and automation (ICRA), Stockholm, pp 1854–1859. https://doi.org/10.1109/icra.2016.7487330

Van Brummelen J, O'Brien M, Gruyer D, Najjaran H (2018) Autonomous vehicle perception: the technology of today and tomorrow. Transp Res Part C: Emerg Technol 89:384–406, ISSN 0968-090X, https://doi.org/10.1016/j.trc.2018.02.012

Neural Networks and Advanced Algorithms for Intelligent Monitoring in Industry

Philipp A. E. Schmid, Alexander Steinecker, Jianwen Sun and Helmut F. Knapp

Abstract How will machines and plants be maintained in the future? Opinions differ widely. Some are propagating industry 4.0 as a key technology, others are very skeptical about the huge data collection needed for this approach. However, a lot of machine data is already available today—it only needs to be used intelligently. Thanks to the latest boost in the field of neural networks empowered by the availability of computing power at low cost, there are great opportunities for condition monitoring and predictive maintenance.

Keywords Deep learning · Predictive maintenance · Machine learning · System monitoring · Industrial processes

1 Introduction

Every plant operator desires a reliable health status report of his assets and if possible a prediction of when the individual sub-systems and components will reach the end of their life. The aim is to ensure long-term operation, detect faults at an early stage and prevent failures. If a machine fails, not only costs for repairs or new purchases count, but also factors such as quality, adherence to delivery dates and maintenance of the production chain. In extreme examples such as a copper mine or oil refinery, the failure of a single machine can paralyze the entire production process. In such cases, the costs for the operator can quickly reach several million Euros per day. Technical maintenance often involves a fine balance between effort, down-time and costs that could be incurred in the event of a system failure.

Industrial plants are already closely monitored today in order to detect or prevent failures of partial or complete systems as early as possible. Today's concepts for planned maintenance can at best limit sudden random failures, but they are slightly inflexible. Either components are replaced after a fixed interval or after a

P. A. E. Schmid · A. Steinecker (✉) · J. Sun · H. F. Knapp
Regional Center Alpnach, CSEM, Untere Gründlistrasse 1, 6055 Alpnach, Switzerland
e-mail: alexander.steinecker@csem.ch

J. Langheim (ed.), *Electronics Components and Systems for Automotive Applications*, Lecture Notes in Mobility,
https://doi.org/10.1007/978-3-030-14156-1_14

fixed mileage, or they are measured for wear and tear and replaced if a control value passes a certain threshold. Both concepts lead to rather high regular maintenance costs.

CSEM is aiming at intelligently using as much as possible the available machine operating parameters for advanced data analytics. Artificial neural networks are the key to this. An artificial neural network has the ability to learn things independently and apply what it has learned to new situations. It is able to detect hidden dependencies and complex correlated patterns. One of the main reasons why such networks are becoming established in industry right now is the inexpensive and compact availability of computing power. A common network with several million synapses or mathematical equations can nowadays be processed on a modern smartphone without any problems.

2 Background

Condition monitoring and prognosis (predictive maintenance) are the processes of assessing the health status of an in-service equipment and suggesting appropriate following operations or maintenances, respectively. Compared to the conventional time-based maintenance, condition based monitoring and prognosis enable timely corrective maintenances and prevent unexpected failures. Various methods and techniques have been developed in this area during the past decade. Most of these methods fall into three categories: model-based approaches, data-driven approaches, and hybrid approaches (Lee et al. 2014). Model-based approaches usually require the data collected from nominal and degraded conditions. If these data are not available, data under different working conditions will be generated from a simulated model of the machine (Luo et al. 2003). The primary disadvantage of these model-based approaches is their dependence on the simulated models. If the simulated model is not accurate, predictions concluded on the basis of the generated data are not reliable anymore. Furthermore, model simulation for each machine is tedious and time consuming, especially if the assessed machine is complicated in structure.

Motivated to overcome the requirement of an accurately simulated model, many data-driven approaches have been developed. Conventional data-driven approaches employ different filters, such as Kalman filters (Wu et al. 2004; Zhan et al. 2006), to predict the remaining life of the critical component through different kinds of extracted features, e.g. the vibration levels, and waveform signatures. Modern data-driven approaches are mostly based on machine learning algorithms, such as fuzzy logic (Perovic et al. 2001; Wang et al. 2004), neural networks (Kan et al. 2015), support vector machines (Kan et al. 2015; Susto et al. 2015; Liu et al. 2017) and others. One drawback of these methods is the effect of selected data features on the accuracy of final predictions. Different kinds of data features are elaborately hand-crafted by researchers to explore the underlying patterns, such as the spectrum, root mean square, kurtosis, etc. Nevertheless, with the increasing complexity of systems investigated, patterns embedded in the data get more and more complicated at the same time. Existing hand-crafted features may not represent the data comprehensively any more.

Hybrid approaches refer to the methods combining model-based and data-driven methods. Some of these hybrid approaches employ data-driven methods to extract data features, making the data-driven method as a preprocessing step for the simulated model (Lee et al. 2014; Kan et al. 2015). On the contrary, some hybrid approaches use explicit domain knowledge based on the model to obtain data features, and feed these obtained features to the data-driven methods for further prognostic (Garga et al. 2001). While inheriting the advantages of model-based and data-drive approaches, hybrid approaches also have similar drawbacks, such as its high demand for expert-knowledge to design the model and select the appropriate feature-extractor.

Through reviewing current predictive maintenance approaches, we find that a main bottleneck of currently available techniques is the selection of data features. How to select the best features becomes a primary question while constructing a predictive maintenance system. An automated selection of features is strongly desired.

In the recent decade, deep learning has become dominant in the machine learning community for its strong feature self-learning ability (Bengio et al. 2013; Goodfellow et al. 2016). It extracts multiple level representations of the data through its deep architecture, and learns the underlying structures and sophisticated high-level features of the data (Bengio et al. 2013; Goodfellow et al. 2016). Deep learning has achieved successes in different domains, such as object recognition (Cireşan et al. 2012; He et al. 2015), natural language processing (Peters et al. 2017; Yu et al. 2017) or speech recognition (Amodei et al. 2016; Xiong et al. 2018). Attracted by its strong feature self-learning ability, we aim to explore how to utilize deep learning algorithms to perform predictive maintenance for industry applications.

3 Application of Deep Learning

Monitoring approaches, such as CSEM is following, relies on processing of hybrid signals. If these data are processed via neural networks faults could be detected before the machine fails. It may even be possible to estimate the remaining time before a defect occurs. CSEM has developed monitoring and predictive maintenance software based on neural networks. The system works in three steps, which are explained in more detail below:

- detecting a deterioration of a machine (detection)
- identification of the components responsible for the malfunction (identification)
- prediction of how the condition of the machine will develop (prediction).

Detection

The neural network learns how a machine behaves in the normal operating range. Depending on the settings, processed parts or ambient conditions, the measured values of a machine can change considerably. Such patterns must be recognized and stored in the network. The more complex a machine is, the more dependencies there are between different sensors and actuators. This is where artificial intelligence beats human beings. They can internalize cross-correlations independently and without expert knowledge thanks to training data. Once the network has learned the default behavior of a machine, it can detect deviations and evaluate such anomalies. Since the artificial network perceives the machine as a whole like an experienced employee, it detects all deviations, not just those previously defined in an error catalog.

Identification

The first two approaches are generic, i.e. universal methods of artificial intelligence. They take place on a superordinate level and show that the system no longer runs normally and how long it takes until it fails completely. However, they cannot identify the cause of the disruption. To do this, the system has to search in the intermediate layers of the neuronal network and make its way back from the brain to the body or machine, so to speak. Which neurons were activated? Which input variables are received by these neurons? Which sensor provides this data? In this way, the neural network can classify the causes, identify the responsible components and help to take appropriate measures.

Prediction

Once the system has recognized that the machine is drifting out of the normal operating range, the next question is in which time horizon the disturbance occurs. Often a linear regression can only achieve basic results here. A more accurate prediction requires a deep understanding of the machine. Not only short-term changes are highly relevant, but also states of the machine that date back longer. In order to integrate such past events, so-called recurrent neural networks have been established. These networks can store relevant information but also forget unnecessary information. With this category of artificial intelligence it is possible to make very adequate predictions of a drifting machine. The forecasts can be adapted to the current situation of the machine every minute, recalculated and evaluated accordingly.

4 Methodology

Among various deep learning algorithms, we select Convolutional Neural Networks (CNNs) as the starting point, based on its huge success in image recognition area and its ability to extract multi-layer feature representations automatically. In this section, we will introduce the system we build at CSEM.

Deep learning basis

As shown in Fig. 1 below, typical CNNs consist of two parts: the feature extraction part composed by convolutional layers and pooling layers, and the classification part formed by fully connected forward networks.

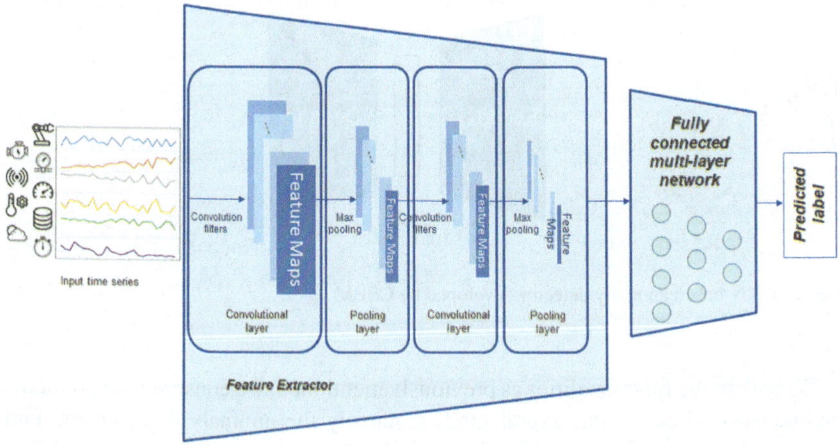

Fig. 1 A typical CNN structure

The corner stone of the feature extractor in a CNN is the working unit assembling convolutional layers and a pooling layer. As indicated by the name, convolutional layers first apply convolution filters to the input data, the time series data or feature map in our case, provided by preceding layers. Afterwards, the convolutional layers apply an activation function, which brings in the nonlinearity, to the convolved input data to generate their output. This operation can be formally expressed as:

$$X_i = f(b_i + W_i * X_{i-1})$$

where X_i is the output of the ith layer (when $i = 1$, X_0 is the input data), W_i is the convolution filter of the ith layer, b_i is the bias vector of the ith layer, and $f(\cdot)$ is the activation function, which can be sigmoid, ReLu, and so on. With such operations, convolutional layers can detect the local conjunctions of features from their input data and produce a feature map.

Pooling layers employ different methods to reduce the dimension of feature maps, aiming to increase the invariance of extracted features as well as to reduce the computational work. Typical pooling methods include max-pooling and sum-pooling.

By concatenating such working units one after another, CNN constructs a feature extractor, which can extract feature maps at different level with different dimensions. The more layers it contains, the more abstract the representations are.

After extracting high-level feature representations, typical CNNs use a multi-layer fully connected forward network to perform the classification task. The number of neurons in the output layer corresponds to the number of classes of the data. The output of each neuron of the output layer indicates the probability of the time series classified as the class represented by that neuron. Typically, we classify the input time series as the class with a maximum probability.

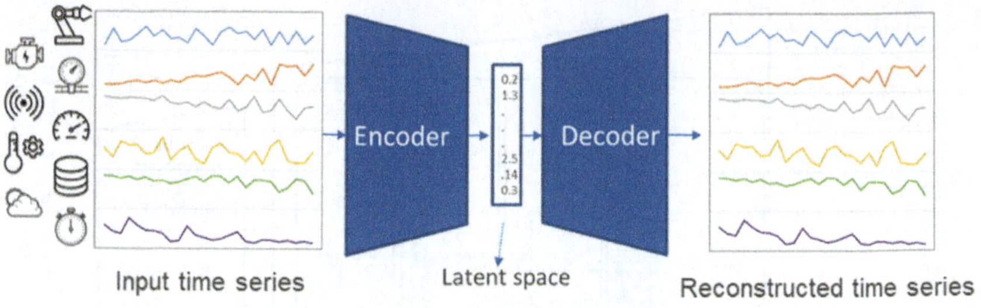

Fig. 2 CNN-based anomaly detector developed by CSEM

To realize the functionalities as previously mentioned, we construct the predictive maintenance system with several models, namely the anomaly detector, the fault identifier, and the fault predictor.

Anomaly detector

Anomalies are the abnormal behaviors of a system. At the production site, there is always a large amount of data from normal working conditions, while data from abnormal working conditions is scarce. Acknowledging this fact, we build the anomaly detector with a CNN-based Auto-Encoder (CNN-AE) that only learns the normal working conditions, because knowing how the system works in normal conditions is sufficient. Every behavior deviating from the normal conditions is considered as an anomaly and will be further analyzed.

As illustrated in Fig. 2, the feature extractor of a CNN acts as the encoder, while the transposed feature extractor acts as the decoder. With such a structure, the CNN-AE learns how to reconstruct the time series as closely as possible. Normally, the latent space, i.e. the high-level feature representations, are narrower than the input layer to avoid the overfitting problem. By feeding data only from normal working conditions, we can train a CNN-AE to learn what good working conditions look like.

Fault identifier

Once an anomaly is detected, we would like to know the possible reasons for such an abnormal behavior. Therefore, we build a fault identifier with a typical CNN.

One important step to train a fault identifier is the dataset preparation. In order to classify the anomalies reliably, it is necessary to prepare a dataset that consists of data not only from normal working conditions but also from commonly seen faulty working conditions, which can be summarized based on the experiences from field service. With a well prepared dataset, we can train a reliable fault identifier to predict the possible cause. If the probability of the anomaly belonging to any pre-defined fault condition is low, we will add it to the databank for further study and fine-tuning of trained models.

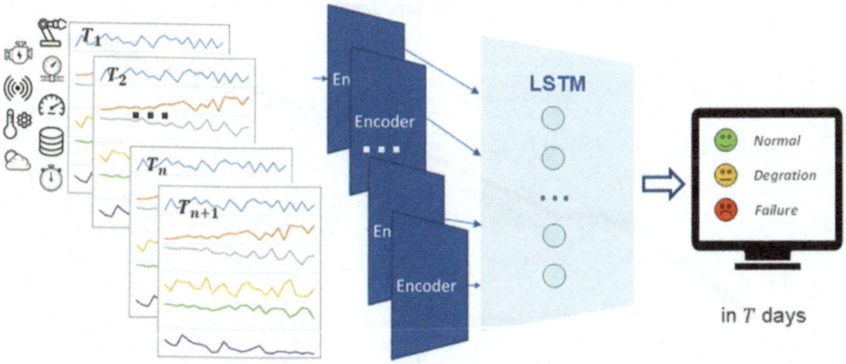

Fig. 3 CNN-based fault predictor developed by CSEM

Fault predictor

To predict possible abnormal working conditions, we not only need the CNN feature extractor to learn high-level feature representations, but also need a Long-Short Term Memory (LSTM), a type of Recurrent Neural Network (RNN), to learn the temporal patterns among the sequential data, so as to perform the prediction task. Figure 3 illustrates the fault predictor developed by CSEM.

The feature extractor of CNNs still acts as an encoder in the fault predictor. Unlike the fault identifier that applies a fully connected multi-layer forward network directly to the extracted latent features, the CNN-based fault predictor concatenates a set of CNN-encoders and applies them to a sequence of time series in chronological order. A sequence of latent features will be extracted and forwarded as the inputs for an LSTM network. The LSTM network then learns the temporal patterns buried within the latent feature sequence to predict the next possible working conditions, i.e. to predict the working state in the next Tth day.

5 Case Study

We tested the algorithms with several different systems. One of the study cases is a complex system producing compressed air. For such a system, we have recorded various kinds of signals, such as the pressure signal, the power signal, the temperature, and so on. We were able to define a list of commonly seen abnormal conditions in the field and collected data for these abnormal conditions.

The dataset was recorded from a test system, using the same signal acquisition system as deployed in the field. For the training and testing, we mainly used the pressure signals and the power signal, as shown in Fig. 4.

In the anomaly detector experiments, we trained a CNN-AE with 70% of the collected signals from good working conditions and tested it with the remaining 30%

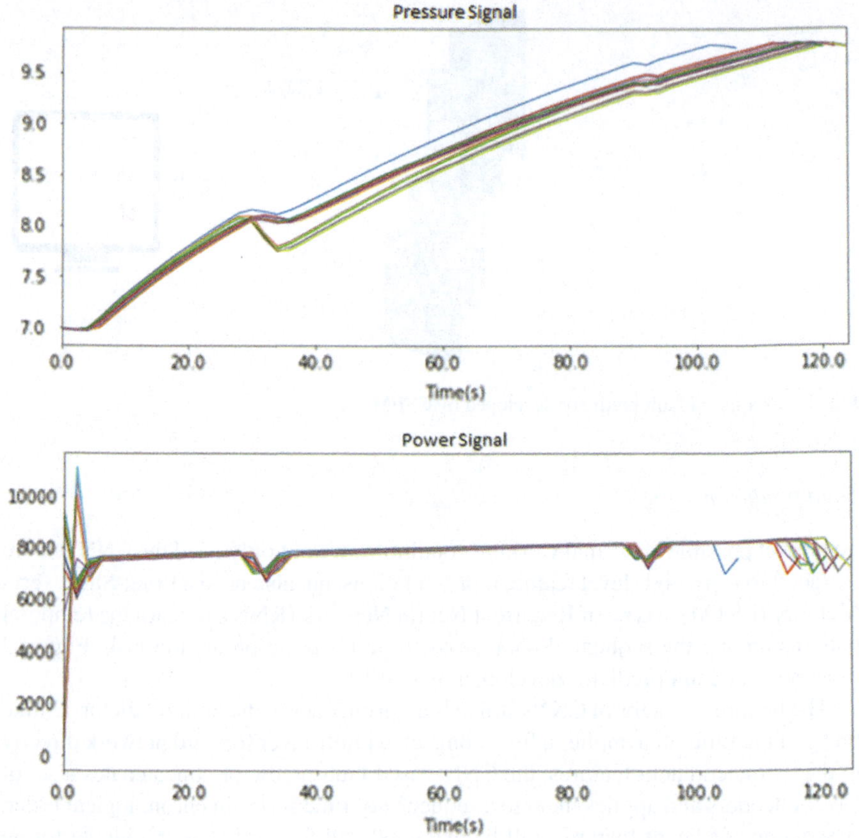

Fig. 4 Example of signals that we recorded from the test system (arbitrary units on y-axis)

signals of good working conditions as well as all the signals from abnormal working conditions. The diagram to the right presents the result in terms of a confusion matrix. As indicated in the table, the trained CNN-AE can achieve an accuracy of 97.5%.

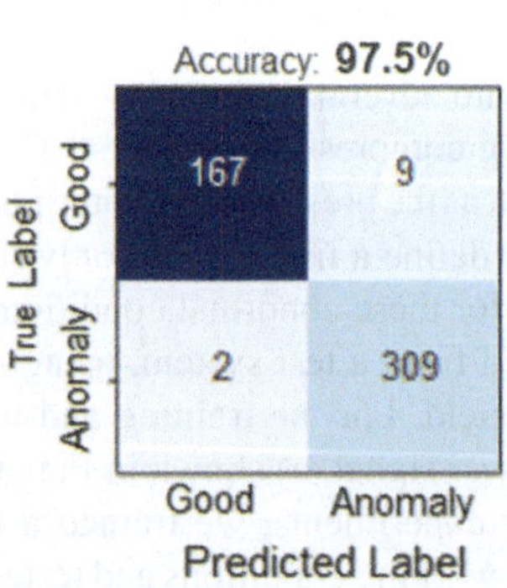

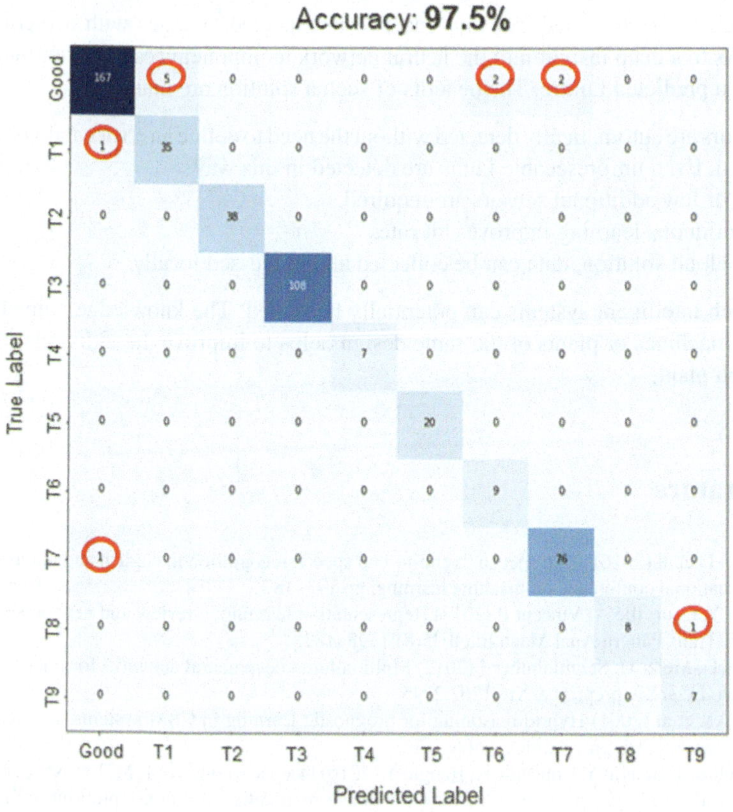

Fig. 5 Result of the fault identifier experiment

In the fault identifier experiments, we trained a CNN again with 70% of the collected signals from all cases and tested it with the remaining 30%. Figure 5 presents the result also with a confusion matrix.

Red circles in Fig. 5 mark the number of misclassified samples of each category. As presented in Fig. 5, the trained fault identifier can achieve an accuracy of 97.5%, which reaches the expectation.

6 Synthesis

Neural network approaches, such as the one CSEM is using, aim to create a brain of a machine. They are able to abstract the system and understand it accordingly. This makes it possible to recognize significant disturbances—even unpredictable ones.

With the available operating data, the approach independently acquires knowledge of the relevant characteristics of the machine. It builds up a wealth of experience through the history and can thus react to unexpected changes with a prediction. Thanks to a deep insight into the neural network, components can be identified that cause a predicted failure. The benefits of such a solution are manifold:

- Errors are automatically detected without the need to define an error catalog beforehand. Even unforeseeable faults are detected in this way.
- No or few additional sensors are required.
- Continuous learning improves hit rate.
- No cloud solution, data can be collected and processed locally.

Such intelligent systems can potentially be scaled. The knowledge gained from other machines or plants of the same design helps to improve the individual results of each plant.

Literature

Amodei D et al (2016) Deep speech 2: end-to-end speech recognition in english and mandarin. In: International conference on machine learning, pp 173–182

Bengio Y, Courville A, Vincent P (2013) Representation learning: a review and new perspectives. IEEE Trans Pattern Anal Mach Intell 35(8):1798–1828

Cireşan D, Meier U, Schmidhuber J (2012) Multi-column deep neural networks for image classification. In: arXiv preprint arXiv:1202.2745

Garga AK et al (2001) Hybrid reasoning for prognostic learning in CBM systems. In: Aerospace conference, IEEE proceedings, vol 6. IEEE

Goodfellow I, Bengio Y, Courville A, Bengio Y (2016) Deep learning, vol 1. MIT press, Cambridge

He K et al (2015) Deep residual learning for image recognition. In: arXiv preprint arXiv:1512.03385

Kan MS, Tan AC, Mathew J (2015) A review on prognostic techniques for non-stationary and non-linear rotating systems. Mech Syst Signal Process 62:1–20

Lee J et al (2014) Prognostics and health management design for rotary machinery systems—reviews, methodology and applications. Mech Syst Signal Process 42(1):314–334

Liu S, Hu Y, Li C, Lu H, Zhang H (2017) Machinery condition prediction based on wavelet and support vector machine. J Intell Manuf 28(4):1045–1055

Luo J, Namburu M, Pattipati K, Qiao L, Kawamoto M, Chigusa S (2003) Model-based prognostic techniques. Update, 1, 2

Perovic S et al (2001) Fuzzy logic system to detect pump faults from motor current spectra. In: Industry applications conference

Peters ME, Ammar W, Bhagavatula C, Power R (2017) Semi-supervised sequence tagging with bidirectional language models. In: arXiv preprint arXiv:1705.00108

Susto GA et al (2015) Machine learning for predictive maintenance: a multiple classifier approach. IEEE Trans Industr Inf 11(3):812–820

Wang W, Farid Golnaraghi M, Ismail F (2004) Prognosis of machine health condition using neuro-fuzzy systems. Mech Syst Signal Process 18(4):813–831

Wu J, Huang C, Huang R (2004) An application of a recursive Kalman filtering algorithm in rotating machinery fault diagnosis. NDT E Int 37(5):411–419

Xiong W, Wu L, Alleva F, Droppo J, Huang X, Stolcke A (2018) The Microsoft 2017 conversational speech recognition system. In: 2018 IEEE international conference on acoustics, speech and signal processing (ICASSP), pp 5934–5938

Yu LC, Wang J, Lai KR, Zhang X (2017) Refining word embeddings for sentiment analysis. In: Proceedings of the 2017 conference on empirical methods in natural language processing, pp 534–539

Zhan Y, Viliam M, Jardine A (2006) Adaptive state detection of gearboxes under varying load conditions based on parametric modelling. Mech Syst Signal Process 20(1):188–221

European Processor Initiative (EPI)—An Approach for a Future Automotive eHPC Semiconductor Platform

Mario Kovač, Dominik Reinhardt, Oliver Jesorsky, Matthias Traub, Jean-Marc Denis and Philippe Notton

Abstract In this paper we present a novel approach for a future automotive embedded high-performance computing (eHPC) platform. The platform is based on an European Processor Initiative (EPI), Common Platform (CP). This paper also gives an overview of the automotive industry's challenges and the general architectural aspects of EPI CP.

Keyword High-performance computing · EPI · European processor initiative · Common platform · Automotive · Vehicle architecture · eHPC · ADAS · AUTOSAR · Automated driving

M. Kovač
Faculty of Electrical Engineering and Computing, University of Zagreb, Unska 3, 10000 Zagreb, Croatia
e-mail: mario.kovac@fer.hr

D. Reinhardt · M. Traub
Research, New Technologies, Innovations, BMW AG, Parkring 19, 85748 Garching Hochbrück, Germany
e-mail: dominik.reinhardt@bmw.de

M. Traub
e-mail: matthias.traub@bmw.de

O. Jesorsky
Business Management, Elektrobit Automotive GmbH, Am Wolfsmantel 46, 91058 Erlangen, Germany
e-mail: oliver.jesorsky@elektrobit.com

J.-M. Denis
Strategy & Plan, BDS Strategy & Innovation, ATOS, Paris, France
e-mail: jean-marc.2.denis@atos.net

P. Notton (✉)
BDS Strategy & Innovation, ATOS, Paris, France
e-mail: philippe.notton@european-processor-initiative.eu

J. Langheim (ed.), *Electronics Components and Systems for Automotive Applications*, Lecture Notes in Mobility,
https://doi.org/10.1007/978-3-030-14156-1_15

1 Introduction

The importance of high-performance computing has been on the rise in the last years, and it is expected that this trend will not only continue but rapidly grow. Industry reports show that annual global IP traffic will soon reach several zettabytes, that vast amounts of new devices collect and store data, that scientists are exploring new computing approaches to solving global challenges, that the industry is changing the way products are designed, and that we, as individuals, are constantly expecting more personalized services for the entire spectrum of our lives' aspects (Gentner and Manz 2017; Meyer et al. 2015; Harnisch 2016; Vermesan 2015).

The need to collect and process the vast amounts of data in an efficiently and timely manner comes at a price. The existing approach to HPC systems design is no longer sustainable for the exascale era in which exascale systems are capable of executing 10^{18} calculations per second). Energy efficiency is of enormous importance for sustainability of future exascale HPC systems.

Europe has recognized this global HPC challenge and has strategically initiated the efforts to support the next generation of computing and data infrastructures. EU efforts are synchronized in the establishment of the EuroHPC Joint Undertaking, a legal and funding entity which will enable pooling of the EU and national resources on HPC to acquire, build, and deploy the most powerful supercomputers of the world in Europe.

The European Processor Initiative (EPI) is one of the cornerstones of this EU HPC-strategic plan. EPI draws 23 partners from ten European countries, with the aim to bring a low power microprocessor to the market and to ensure that the key competence of high-end chip design remains in Europe (Kovač 2018). Thanks to these new European technologies, European scientists and industry will be able to access exceptional levels of energy-efficient computing performance. This EPI efforts will benefit Europe's scientific leadership, industrial competitiveness, engineering skills and know-how, and the society as a whole.

One of EPI's core activities will take place in the automotive sector, providing architectural solutions for a novel embedded high-performance computing (eHPC) platform and also ensuring the overall economic viability of the initiative.

EPI's automotive stream activities are targeting main trends driving the innovations in automotive industry including the introduction of autonomous driving (Levels 4 and 5) and the Connected Car infrastructure. New autonomous vehicle network architectures require computing platforms to be able to execute complex vehicle perception algorithms that include sensor/imaging processing, data fusion, environment sensing and modelling, low-latency deep machine learning for object classification and behavior prediction with seamless, dependable and secure interaction between mobile high-performance embedded computing, and stationary server-based high-performance computing.

2 Future Challenges for the Automotive Industry

The European automotive sector with its original equipment manufacturers and suppliers represents an important share on the global market. In nearly every European country the automotive industry is one of the most important domains for the national economy. Meanwhile, the automotive domain is characterized by high complexity and ICT-driven innovation with steering and control of most of the automotive functions by performant embedded computing units (Electronic Control Unit—ECU) (Gentner and Manz 2017; Meyer et al. 2015; Harnisch 2016).

Applications in the automotive sector have to fulfil highly challenging environmental requirements, endure a typical temperature range from −40 to 125 °C and above, as well as mechanical stress and chemical and moisture resistance over a lifetime of at least 15 years. Most of the automotive functions are real-time-relevant and have to fulfil not only functional safety but also cyber-security requirements (Meyer and Gliwa 2011; Bradatsch et al. 2011).

The development in the automotive domain is currently characterized by three main trends:

- The transition from internal combustion engines to full electric vehicles,
- The introduction of highly complex and safety-relevant ADAS assistant systems including first automated driving (AD) functions which demands fail-operational systems, and
- The ‘Connected Car’, its integration into large-scale traffic management systems, and combining both high-performance embedded computing platforms in cars and exascale-class computing centers.

As a result, automotive innovation trends require significantly increased computing performance. The current vehicle network architectures with up to 100 dedicated ECUs are no longer capable to answer the requirements towards increased computing performance, fail-operational and functional safety, cyber-security, and real-time behavior (RT). To reduce the number of ECUs, large scale software integration (LSSI) is aimed to consolidate smaller ECUs on single integration platforms (Fig. 1, Vermesan 2015).

As of today, embedded automotive-qualified processors are not powerful enough to handle issues like sensor data fusion from e.g. camera, radar, and lidar. At the moment, the automotive industry is addressing this by using high-performance CPUs and GPUs from other application domains like e.g. NVIDIA and Intel chips from the consumer and performance computer industry in combination with automotive-qualified master ECUs like e.g. the Infineon AURIX.

Specific automotive processors (number crunchers) adapted from the HPC-domain in combination with qualified innovative automotive MCUs could solve the problem. From the system viewpoint, adapting and using consumer processors must under no circumstances have a negative impact on the function, the real-time behavior, and the availability and the reliability of the automotive compute platform.

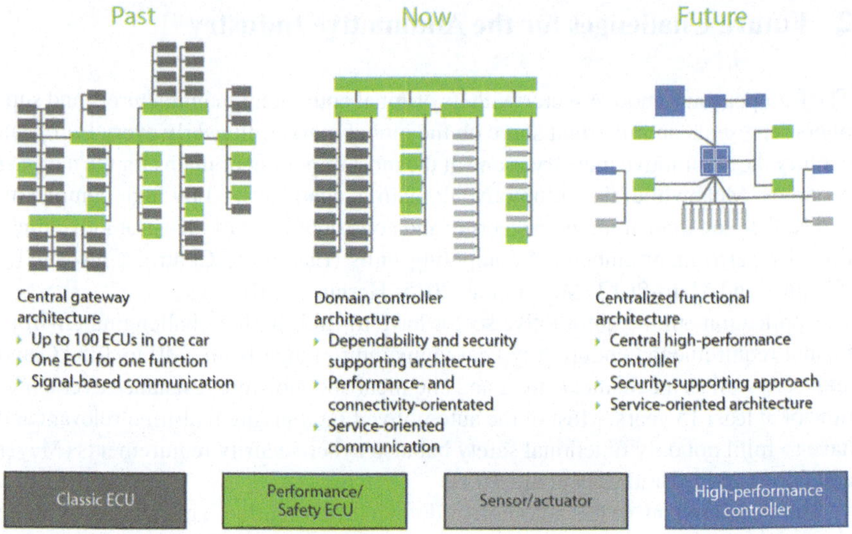

Fig. 1 Consolidation of automotive ECUs and functionality towards a centralized function architecture

Future vehicle network architectures that are based on only a few powerful high-performance computing platforms and that are capable to implement vehicle perception tasks in real time in a fail-operational manner are highly required (Fig. 1).

Looking to autonomous driving functionality vehicle perception for building a model of the surrounding environment is computationally intensive. It includes sensor processing, computer vision, data fusion, object classification, and behavior prediction. Furthermore, the borders for vehicle network functionality are not fixed and the automotive domains are merging, dissolving, or getting into background for new innovations. Car functionality will be classified in different classes of ECUs (Traub et al. 2017). OEM-specific innovations are integrated onto high-performance integration ECUs. These are interconnected by a fast fieldbus like automotive Ethernet.

The expected performance levels for future autonomous vehicles with regards to data produced by the sensors in the vehicles and the bandwidth requirements so the vehicle can properly interact with its environment have been analyzed (Gartner 2018) and are shown in Table 1. It can be observed that high performance low power processors and accelerators are a must for Autonomous Driving (AD) Level 4 and Level 5 vehicles.

The anticipated evolution of processing capacity is large (×2) from Level 3 to Level 4 vehicles and from Level 4 to Level 5 (×1.5). In the very classical HPC world, this evolution of performance would be manageable, considering the significant evolution of the electrical power consumption too. But, embedded compute elements have totally different constrains because of the limited embedded electrical power.

Thus, a simplistic approach exclusively based on an evolution of the compute power cannot match the target performance in a reasonable electrical envelope. As

Table 1 From AD Level 3 to Level 5 vehicles: evolution of processing capacity, storage and data link interface

System parameters	Level		
	3	4	5
Processing capacity (TFLOPS)	40	80	120
DRAM (GB)	1	10	20
Non-volatile storage (GB)	100	300	500
Data link interface (Gbps)	0,1	1	1

opposed to this "*brute force*" approach, the architecture concepts described hereunder will fill the gap between the TFLOPS (10^{18} floating point operations per second) requirements and the electrical power a vehicle can provide.

3 EPI Common Platform Approach

Exascale computation systems will need to simultaneously meet challenges related to performance, system cost, and energy efficiency. To deliver performance, a vast amount of resources is required, but the wrong choices of components, architecture, or implementation might result in a system which is much too expensive and too power-hungry. To find the right balance, global system level optimization is necessary.

EPI therefore intends to share a strong set of common technology between different application domains. Starting from the selection of cutting-edge processor technology, a low-power design approach will be centralized in the hardware around massive parallelism, specialized architecture, low-voltage operating point and fine grain power management. The software stack will be designed to integrate and take advantage of these features to achieve high-energy efficiency and maximize performance across a wide range of layers from the low level firmware, all the way up to system software and application run-times. For this purpose, EPI will harmonize the heterogeneous computing environment by defining a common approach: the so-called **Common Platform (CP)**. It will include the global architecture (hardware and software) specification, common design methodology, and global approach for power management and security.

The CP is organized around a 2D-mesh Network-on-Chip (NoC) connecting computing tiles based on a general-purpose CPU core with built-in FPU acceleration. RISC-V based high energy efficiency accelerators, MPPA accelerators or any other application-specific accelerator.

A common software environment between heterogeneous computing tiles will harmonize the system as well as acting as a common backbone of IP components for IO connection with the external environment such as memories and interconnected or loosely coupled accelerators. With this CP approach, EPI will provide an environment that seamlessly integrates any computing tile. The right balance of computing resources for application matching will be defined through the ratio of accelerator and general-purpose tiles.

4 Concepts for Automotive HPC Platforms Within EPI

One of the key challenges of EPI is to explore and validate the scalability and deployment of computation concepts from HPC down to embedded application domains like automotive. In the following, we describe our methodology for automotive eHPC hardware and software codesign.

4.1 Requirements for the Automotive eHPC

In order to focus on sustainable concepts for embedded computing platforms capable of real-time processing, it is evidently necessary to improve architecture, connectivity, and computing performance of automotive-qualified multi-cores. The outcome of the EPI in this area should be a performant fail-operational automotive real-time multi-core processor.

4.1.1 Processing Power

Future emerging automotive applications raise issues like big data handling, deep machine learning, and combined on-board processing with server-based offloading, which will strongly rely on a seamless and dependable interaction between mobile high-performance embedded computing and stationary server-based high-performance computing. From this perspective, a lot of similarities are visible and therefore opportunities for collaboration between automotive and HPC in terms of architecture, hardware, and software arise. Within the scope of the EPI, it is the intention to explore and validate the scalability and deployment of computation concepts from HPC down to embedded application domains like the automotive domain.

4.1.2 Real-Time Performance

In order to focus on sustainable concepts for embedded computing platforms capable of real-time processing, it is also evident and necessary to improve the architecture, connectivity, and computing performance of automotive-qualified multi-cores. The outcome of the EPI in this area will be a performant fail-operational automotive real-time multi-core processor.

Apart from the automotive domain, this type of platform might be used as a reference implementation for other high-performance embedded computing domains like industry, machinery, and healthcare.

4.1.3 Environmental Settings

The intended system Automotive eHPC Compute Platform must operate in a typical ambient temperature range from −40 to 125 °C and beyond. In addition, defined usage profiles have to be observed. In any case, the automotive-qualified processors (eHPC-MCU) must be able to monitor the correct function of the slave processors. They must safeguard the real-time capability, protect the system for security reasons, and perform a reduced application in parallel in order to take over in case of unavailability or degradation of the HPC GPP.

4.1.4 Energy Efficiency

The trend for future vehicles within the next decade is strongly pushing towards electrical cars or hybrid variants. Therefore, energy-efficient chips are necessary to fulfill the previously mentioned environmental requirements for heat dissemination, chip robustness, and reliability. But similar to smartphones, the computational outcome (MIPS per Watt) must be taken into account for future HPC platforms. The goal should be to extend the electric operation and the resulting driving range (battery-driven) for as long as possible.

4.1.5 Safety

Autonomous driving is mentioned as one of the most customer-intensive game changer of the next decade. This freedom brings new requirements for safety and security. The driver will not be the main controlling instance in severe traffic situations. As a consequence, the eHPC computing platform has to achieve fail-operational requirements in order to be suitable for VDA (German Association of the Automotive Industry) Autonomous Driving Level 4 and 5.

4.1.6 Security

Because cars have typically a long product lifecycle time, the possibility for updates and even upgrades are necessary in our next car generations. Nowadays, security mechanisms can be obsolete in five or more years. Computing power will rise more and more and today's encryption mechanisms could be broken in the future. Looking to quantum computing, even new technologies must be taken into account, which brings new possibilities and dangers in parallel.

4.2 Hardware Platform for the Automotive eHPC

Without innovative solutions, the digital progress in the automotive sector will end in a deadlock because of insufficient computing power for new and increasing fields of application like 360° environment recognition and other real-time systems.

In order to continue on the road of success and to overcome this dilemma, the basic approach is the use of suitable high-performance processors from the HPC sector. The specific challenge is the integration of those high-performance consumer processors that have to be compliant with the relevant application and environmental requirements in the real-time domain of automotive.

A powerful data fusion platform (Reinhardt et al. 2019), such as the automotive embedded HPC platform, could then be obtained by combining such automotive-certified real-time multi-core processor with adapted HPC-General Purpose processors and accelerators. Apart from the automotive domain, such an eHPC platform might be used as reference implementation for other safety-oriented, eHPC-computing domains like industry, machinery, and healthcare.

Necessary improvements and extensions of the automotive multi-core architecture and fast interfaces to the accelerator and the high-performance computing general purpose processor (HPC-GPP) have to be considered as well as suitable software concepts and complex automotive application scenarios using on-board and off-board computation (e.g. traffic management, AD systems with machine learning). This type of interaction will require reliable and secure communication channels, proper identity management and assurance, while providing adequate data and identity privacy. Therefore, it is necessary to integrate hardware/software co-design but also safety/security co-design to successfully implement an automotive eHPC platform.

The favored fundamental approach is a master-slave constellation with one or several automotive-qualified processors operating as masters and one or several HPC-GPPs and accelerators operating as slaves.

With respect to the HPC domain, one of the main incentive reasons to adapt the HPC-GPP—not only for the original purpose with local servers and super computers but also for embedded domains like automotive—is the expected improved exploitation of the highly expensive development of such a complex GPP.

From the automotive viewpoint, the expected additional computing power should not exceed given cost and energy budgets.

A proposed subsequent integration of the HPC-GPP and HPC accelerators into the automotive eHPC is supposed to be technologically, functionally, and economically successful. In order to ensure this, both from the perspective of the HPC world and the automotive point of view, a couple of fundamental questions must be determined and clarified right from the beginning.

4.3 Software Platform for the Automotive eHPC

As described in the previous chapter, a master-slave constellation with one or several automotive-qualified processors operating as masters and one or several HPC-GPPs and accelerators operating as slaves is targeted.

In order to make best possible use of this powerful architecture, the corresponding software requires a more flexible and suitable architecture, which can represent these dynamics.

For EPI, a Classic AUTOSAR for the master controller will be supplemented with an Adaptive AUTOSAR platform running on the HPC-GPPs. This enables a dynamic software configuration and, with service-based communication and heterogeneous computation, it provides mechanisms that ensure the necessary performance (Pallierer and Schmelz 2017). The hardware resources of the performance cores are separated by a hypervisor. The hypervisor virtualizes the hardware and, in so doing, provides the partitions as virtual machines. In this way, the various Adaptive AUTOSAR and, if required, also Classic AUTOSAR partitions can be created.

The combination of a Classic AUTOSAR system with Adaptive AUTOSAR is a state-of-the-art approach to tackle the combination of performance and security requirements of the vehicle network architecture for technologies like autonomous driving (Grave 2017). Figure 2 provides a sample configuration of a future automotive multi-core performance controller.

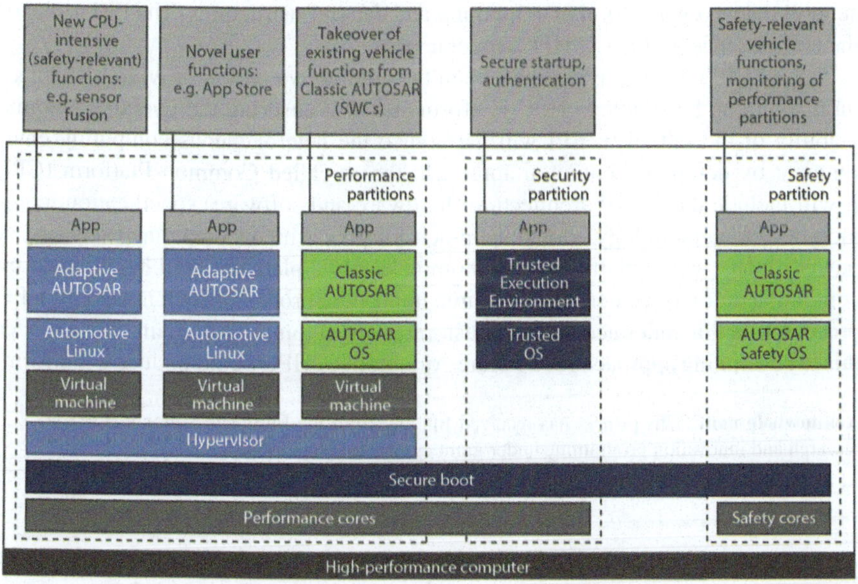

Fig. 2 Architecture of a central control unit with performance and safety cores

The performance of the EPI eHPC platform will be measured with real-world automated driving use cases. Functional modules with high computational demands, such as sensor data fusion and AI components, will be executed in EB robinos (Ohl 2017), a software framework for automated driving which is running on top of Adaptive AUTOSAR.

Code optimization of the functional modules will guarantee the best possible use of the specific EPI platform capabilities, e.g. accelerators. Measurement results will be benchmarked with already existing automotive high-performance controllers.

5 Summary

The European automotive sector with its original equipment manufacturers and suppliers represents an important share on the global market. The automotive domain is characterized by high complexity and ICT-driven innovation with steering and control of most of the automotive functions by performant embedded computing units. In order to focus on sustainable concepts for embedded computing platforms capable of real-time processing, it is evidently necessary to improve architecture, connectivity, and computing performance of automotive qualified multi-cores.

The importance of general high-performance computing has been on the rise in the last years, and it is expected that this trend will continue to grow. Europe has recognized this global HPC challenge and has strategically initiated the efforts to support the next generation of computing and data infrastructures. The EPI is one of the cornerstones of this EU HPC-strategic plan.

One of EPI's core activities will be in the automotive sector, providing architectural solutions for a novel eHPC platform and also ensuring the overall economic viability of the initiative. EPI will harmonize the heterogeneous computing environment by defining a common approach: the so-called Common Platform (CP). It will include the global architecture (hardware and software) specification, common design methodology, global approach for power management and security. A powerful data fusion platform, the automotive eHPC platform, will be obtained by combining automotive-certified real-time multi-core processor with novel EPI HPC processor that includes accelerators customized to automotive domain. We consider this to be winning approach for a Future Automotive eHPC Semiconductor Platform.

Acknowledgment This project has received funding from the European Union's Horizon 2020 research and innovation programme under grant agreement No 826647.

References

Bradatsch C, Ungerer T, Zalman R, Lajtkep A (2011) Towards runtime testing in automotive embedded systems. In: Proceedings of the 6th IEEE international symposium on industrial embedded systems (SIES), pp 55–58

Gartner (2018) Hype cycle for automotive electronics, 2018. https://www.gartner.com/doc/3883996/hype-cycle-automotive-electronics, 25 July 2018

Gentner H, Manz M (2017) Environment model for autonomous driving—the upcoming challenges. ConCarExpo Conference, BMW AG, Berlin

Grave R (2017) The vehicle architecture of automated driving. EB tech paper. https://www.elektrobit.com/tech-corner

Harnisch J, Schneider R, Kohn A, Schmidt K, Schoenberg S, Zhou Q (2014) Efficient virtualisation for functional integration on modern microcontrollers in safety-relevant domains. Article, SAE Technical Papers

Kovač M (2018) Changing approaches to HPC systems. CIO applications Europe. https://www.cioapplicationseurope.com/cxoinsights/changing-approaches-tohpc-systems-nid-520.html

Meyer A, Gliwa P (2011) Measurement of application performance on Automotive microprocessors. VDI Wissensforum "Elektronik im Kraftfahrzeug, Baden-Baden

Meyer A, Schmidt K, Marx D, Harnisch J, Christlbauer H (2015) Non-intrusive tracing at first instruction. In: SAE 2015 world congress & exhibition. https://doi.org/10.4271/2015-01-0176

Ohl S (2017) Staying in lane on highways with EB robinos. EB tech paper. https://www.elektrobit.com/tech-corner

Pallierer R, Schmelz B (2017) Adaptive AUTOSAR for high-performance in-car computer. EB tech paper. https://www.elektrobit.com/tech-corner

Reinhardt D, Dannebaum U, Scheffer M, Traub M (2019) High performance processor architecture for automotive large scaled integrated systems within the European Processor Initiative Research Project. SAE Technical Paper 2019-01-0118

Traub M, Maier A, Barbehön K (2017) Future automotive architecture and the impact of IT trends. J Intell Syst 0740-7459/17, IEEE

Vermesan O, Sans M, Hank P, Farrall G (2015) Advanced electronic architecture design for next electric vehicle generation. Lecture notes in mobility. ISSN 2196-5544

[illegible] with [illegible]

[illegible] (2015) [illegible]

[illegible] (2018) [illegible]

Model-based Schedule Synthesis in Time-Sensitive Networks

Morteza Hashemi Farzaneh and Alois Knoll

Abstract Over the last two decades, various proprietary modifications have been applied to Ethernet technology to enable deterministic communication required for time-critical systems. Time-Sensitive Networking (TSN) standards are under development by the Institute of Electrical and Electronic Engineers (IEEE) and address hard timing requirements of Ethernet-based distributed applications. The main objective of these standards is to offer an open and vendor-independent networking infrastructure for mixed-critical applications. The support of mixed-criticality reduces heterogeneity of networking technologies and simplifies not only the development but also the integration of distributed applications. However, the scheduling and verification of these networks require advanced expertise and are time-consuming. In this paper, a graphical modeling framework is introduced which facilitates scheduling synthesis and configuration of TSN networks. The created graphical models are automatically translated into a network knowledge base, which is the core of the framework. It is used to process queries regarding infrastructural verification and selective information extraction in order to build scheduling constraints.

Keywords Time-Sensitive Networking · Real-time communication · Modeling · Network configuration

1 Introduction

Time-Sensitive Networking (TSN) is a set of new standards which are in developing process by Institute of Electrical and Electronics Engineers (IEEE) to support mixed criticality based on Ethernet technology. These standards will help to overcome challenging requirements of automotive domain considering upcoming innovative

M. Hashemi Farzaneh (✉) · A. Knoll
Technical University of Munich, Munich, Germany
e-mail: hashemif@in.tum.de

A. Knoll
e-mail: knoll@in.tum.de

J. Langheim (ed.), *Electronics Components and Systems for Automotive Applications*, Lecture Notes in Mobility,
https://doi.org/10.1007/978-3-030-14156-1_16

applications such as autonomous driving and infotainment that require e.g. fully deterministic network behavior, high bandwidth, fail-operational and etc.

Despite the strengths of TSN, increasing number of involved sensors, actuators and Electronic Control Units (ECU) and mutual influence of requirements still cause high network engineering overhead for automotive network engineers. For example, considering an existing in-vehicle network and its applications' requirements, reviewing the whole configuration is required when new critical nodes join the network. After reviewing the TSN standards, some parts of the network configuration may need to be modified and after this reconfiguration one has to formally verify that all requirements are satisfied.

To guarantee hard real-time requirements, Time-Aware Shaper (TAS) standard (IEEE 802.1Qbv) is introduced based on the theory of time-triggered communication. Using TAS and based on the frames' priority, frame queuing delays in the egress ports are minimized. Gate Control List (GCL) is a main component of TAS and contains a list of gate opening times for all 8 priorities queues based on IEEE 802.1Q. Each network port contains a GCL and it requires to be correctly configured in order to give latency and jitter guarantees and avoid deadline exceeding.

Finding a feasible time-triggered schedule is a known NP-complete problem (Craciunas and Oliver 2016). To deal with this problem, timing requirements are transformed to schedule constraints and prepared for constraint solvers such as Satisfiability Modulo Theories (SMT) (Steiner 2010; Craciunas and Oliver 2016) or Integer Linear Programming (ILP) (Zhang et al. 2014) to find a feasible schedule for GCL configuration. To achieve this transformation, information about e.g. frames and affected egress ports in relation to their timing requirements have to be extracted.

However, without an adequate network model, this transformation is time-consuming especially at network design time when repeated changes are expected. It also requires expertise in the solver platform in order to obtain a correct transformation. In this paper, we present the developed graphical modeling tool.

2 Related Work

Declarative networking is proposed in (Loo et al. 2009), based on logic programming language Datalog which is a subset of Prolog. A practical declarative network management approach based on Datalog is presented in (Hinrichs et al. 2009). This paper only describes how to specify QoS requirements such as latency, jitter and bandwidth but it does not explain how to use it for configuration and verification.

Simulation-based approaches (Lim et al. 2011; Alderisi et al. 2012; Steinbach et al. 2012; Meyer et al. 2013; Thiele et al. 2015) are developed to analyze the performance of Audio Video Bridging (AVB) and TSN. The main disadvantage of these approaches is that not all of the corner cases can be covered by simulations and therefore are not suitable for required formal verification of critical requirements. In contrast, formal analysis methods are developed in (Thangamuthu et al. 2015; Farzaneh et al. 2016) to verify the performance of the TSN shapers. These methods however, focus only on timing and latency aspects on the network layer and do not respond to the question of mutual influence of critical requirements.

Simulation-based approaches are used to analyze the performance of Audio Video Bridging (AVB) and time-triggered Ethernet. They are very useful to gain an insight in these technologies. However, they cannot cover all corner cases and therefore are not suitable for formal verification of critical requirements. To overcome this problem, formal analysis methods are developed to verify the performance of the TSN shapers. Both simulation-based and formal methods do not deal with configuration challenges of TSN such as schedule synthesis. The configuration and infrastructural verification of TSN networks are discussed in (Farzaneh et al. 2016; Farzaneh et al. 2017; Farzaneh and Knoll 2016) providing a roadmap to enable plug and play functionalities.

3 Requirements

In this section, we introduce a list of requirements which have to be fulfilled by a TSN modeling and configuration framework.

Easy network modeling: to design network topologies, components, application's dataflow, timing requirements, and etc., graphical and textual modeling elements are required which can assist the network designers. Such features are applied to model the network by drag and drop of modeling elements. Software libraries and environments are required which assist to overcome this challenge.

Simplified access to the elements of a created network model: after a network model is created, features are required which allow easy access to the modeling elements by queries. The main intension is to find a solution which reduces the need for exhaustive programming of routines (e.g. using imperative or object-oriented languages) for each specific query.

Selective representation of model data: a network information management feature is required that makes it possible to extract desired information efficiently for configuration and verification tasks. The efficiency means here reducing the overhead of imperative routine programing by following a declarative approach (which allows shorter programs and avoids side effects because of referential transparency).

Logical inference: to use the available network information for verification of desired network properties and also to describe topological or application relations in the network, an inference engine is required which allows to derive logical proofs and to find desired unknowns (via queries).

Constraint-based problem formulation and solving: an environment is required where infrastructural and scheduling constraints can be formulated and solved. The results are used for configuration of network components (TSN switches).

4 Approach

As depicted in Fig. 1, the modeling approach consists of a metamodel, which is used as a basis for the implementation of the graphical modeling tool (Farzaneh et al. 2017) (to created graphical network model instances). Application developers or TSN managers use the graphical editor to specify the dataflow and to annotate the QoS requirements of the developing real-time applications. For example, it can be annotated that a certain application contains periodic data transmission with hard timing constraints or it has reliability requirements and therefore requires e.g. multiple disjoint paths (cf. IEEE 802.1CB) in the network topology. The modeling tool is also used to define the physical properties of the components. Using a formal metamodel based on logic programming and following our previous work (Farzaneh et al. 2016), the graphical model instances are transformed into a knowledge base. We apply Prolog as a concrete implementation of logic programming and transform the graphical network model instance into Prolog facts. Moreover, we implemented Prolog rules to exploit the advantages of logical inference to query and verify interesting network properties such as required redundant disjoint paths. It is also used to query the required information from the network model to build time-triggered schedule constraints.

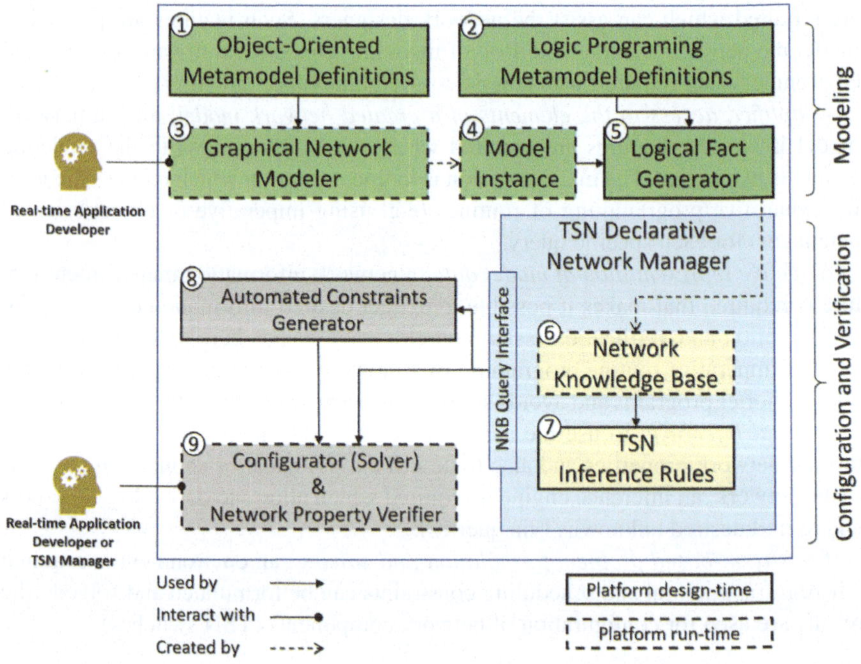

Fig. 1 Framework overview

Based on a Satisfiability Modulo Theories (SMT) solver, the constraints are solved for the decision of whether there is a feasible schedule for the created model instance. The constraints can also be used for verification of manually created schedules. The solver checks the schedule and decides whether all constraints (real-time requirements) are fulfilled.

To define the object-oriented model of the network, we use the Eclipse Modeling Framework (EMF). As described in (Farzaneh et al. 2017), a network model consists of a *network view*, which contains all graphical network components. A network *node* is either a *switch* or an *end-station*. Nodes have at least one Ethernet *port*. Switches and end-stations are connected through ports. A *link* connects two nodes and has exactly one *source* and one *target* port.

A real-time application is presented as a *domain*. Each domain contains one or more data *topics*, which are *published* and *consumed* through ports within a domain. We distinguish between a periodic *topic* and a *sporadic topic*. The periodicity of the topics is highly relevant for schedule synthesis. In contrast to periodic topics, the sporadic topics have an attribute *minInterval* for minimum time between two consecutive frames. This parameter is important for worst-case latency analysis of a sporadic frame. Data topics with tight timing requirements have to be periodic in order to be considered in the schedule synthesis constraints in IEEE 802.1Qbv. Each topic may contain *QoS* requirements such as timing and *fail-operational* to specify e.g. the number of required redundant links.

5 Implementation and Case Study

The objectives of the case study are to demonstrate the capability of the graphical modeling tool to synthesize a correct schedule for TAS. Moreover, we show how it gives assisting feedback to developers and network managers to change their model in order to correct its design problems and achieve a correct synthesized schedule.

We designed an exemplary network model, which is presented in Fig. 2. Network designers use the modeling elements and build the desired model via drag and drop. They are not confronted with the complex details of generating the scheduling constraints and can focus on high-level network management and application developments tasks.

Two switches, which eight ports each connect end-stations that are physically separated by the switches. Two domains are available which describe the communication topics. Using the network knowledge base, we extracted the 14 streams carrying topics through the network. The details about streams, topics, non-overlapping relations of the streams, and the synthesized schedule are available in Fig. 3.

The stream's offset values are all in a correct range between zero and the streams' periods. Streams s_2, s_3, s_7, s_10, s_14 do not have non-overlapping requirements and thus, have the same gate opening time t = 0. To implement this experimental case study, we used a desktop PC with an Intel Core i7 CPU and 4 GB RAM. For the synthesis, we used the SMT solver Z3 Version 4.5.0. We use Gigabit Ethernet in our models.

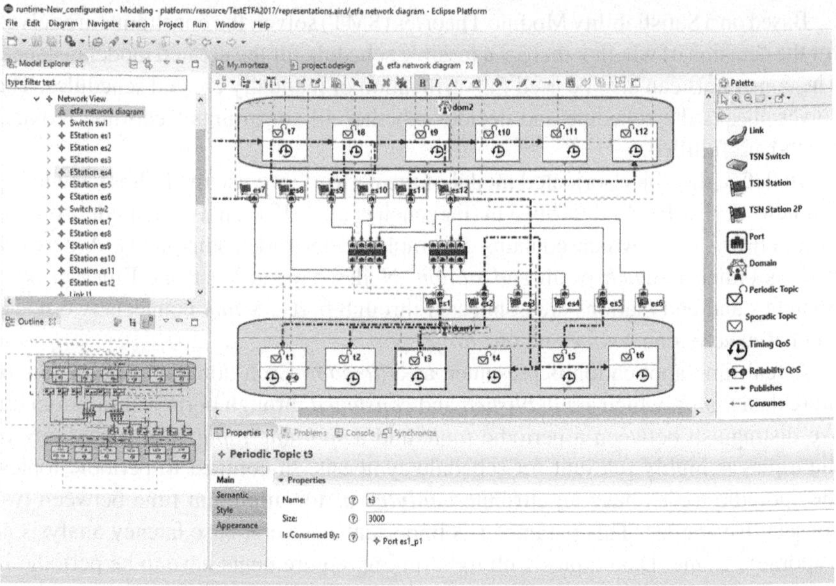

Fig. 2 The graphical modeling tool with adequate modeling elements

Stream	Non-overlapping with	Domain	Topic	Publisher	Consumer	Period (ns)	Length (Byte)	First Gate Opening Time (ns)
s_1	s_2	*dom1*	t_1	$es_1\_p_1$	$es_2\_p_1$	500.000	400	38.072
s_2	s_1	*dom1*	t_2	$es_1\_p_1$	$es_2\_p_1$	2.000.000	1000	0
s_3	s_4, s_5, s_6	*dom1*	t_3	$es_2\_p_1$	$es_1\_p_1$	8.000.000	3000	0
s_4	s_3	*dom1*	t_4	$es_2\_p_1$	$es_3\_p_1$	16.000.000	6000	8.103.096
s_5	s_3, s_7, s_8, s_6	*dom2*	t_7	$es_7\_p_1$	$es_1\_p_1$	500.000	400	186.480
s_6	s_{13}, s_3, s_5, s_{12}	*dom2*	t_{12}	$es_{11}\_p_1$	$es_1\_p_1$	100.000.000	5000	7.705.352
s_7	s_8, s_5	*dom2*	t_7	$es_7\_p_1$	$es_8\_p_1$	500.000	400	0
s_8	$s_5, s_7, s_9, s_{10}, s_{12}, s_{13}$	*dom2*	t_7	$es_7\_p_1$	$es_{12}\_p_1$	500.000	400	167.608
s_9	$s_8, s_{14}, s_{10}, s_{11}, s_{12}, s_{13}$	*dom2*	t_8	$es_8\_p_1$	$es_{12}\_p_1$	500.000	400	186.480
s_{10}	$s_8, s_9, s_{11}, s_{12}, s_{13}$	*dom2*	t_9	$es_9\_p_1$	$es_{12}\_p_1$	5.000.000	5000	0
s_{11}	$s_8, s_9, s_{10}, s_{12}, s_{13}$	*dom2*	t_{10}	$es_9\_p_1$	$es_{12}\_p_1$	100.000.000	10000	99.205.352
s_{12}	$s_{11}, s_{10}, s_9, s_8, s_{13}, s_6$	*dom2*	t_{11}	$es_{11}\_p_1$	$es_{12}\_p_1$	100.000.000	5000	205.352
s_{13}	$s_6, s_8, s_9, s_{10}, s_{11}, s_{12}$	*dom2*	t_{12}	$es_{11}\_p_1$	$es_{12}\_p_1$	100.000.000	5000	481.128
s_{14}	s_9	*dom2*	t_8	$es_8\_p_1$	$es_7\_p_1$	500.000	400	0

Fig. 3 Synthesis results of the case study network model

6 Summary

To automate the process of schedule synthesis in Time-Sensitive Networks, an innovative graphical modeling tool is introduced. It exploits multiple paradigm such as object-oriented modeling, logic programming, and Satisfiability Modulo Theories. However, it is possible that the generated network models are non-schedulable. In this paper, we use the unsatisfiable cores in SMT solvers (generated for models with conflicting constraints), to construct helpful feedback to correct non-schedulable network models step-by-step to find a feasible schedule. We only considered the non-overlapping requirements of time-triggered streams in this paper. The demonstration emphasizes that significant parts of the manual TSN configuration and verification can and must be automated using the developed framework in order to accelerate the TSN-based application developments.

The future work will focus on a plan to extend the modeling tool in order to increase the precision of the models by covering more network details such as application level constraints. Moreover, we intend to formulate and generate constraints for peristaltic shaper of TSN (described in IEEE 802.1Qch—Cyclic Queuing and Forwarding) using our graphical modeling tool.

Literature

Alderisi G, Caltabiano A, Vasta G, Iannizzotto G, Steinbach T, Bello LL (2012) Simulative assessments of IEEE 802.1 Ethernet AVB and time-triggered Ethernet for advanced driver assistance systems and in-car infotainment. In: Vehicular networking conference (VNC). IEEE, pp 187–194

Craciunas SS, Oliver RS (2016) Combined task-and network-level scheduling for distributed time-triggered systems. Real-Time Syst 52(2):161–200

Farzaneh MH, Knoll A (2016) An ontology-based plug-and-play approach for in-vehicle time-sensitive networking (TSN). In: IEEE 7th annual information technology, electronics and mobile communication conference (IEMCON). IEEE

Farzaneh MH, Shafaei S, Knoll A (2016) Formally verifiable modeling of in-vehicle time-sensitive networks (TSN) based on logic programming. In: Vehicular networking conference (VNC). IEEE

Farzaneh MH, Kugele S, Knoll A (2017) A graphical modeling tool supporting automated schedule synthesis for time-sensitive networking. In: 22nd IEEE international conference on emerging technologies and factory automation (ETFA). IEEE

Hinrichs TL, Gude NS, Casado M, Mitchell JC, Shenker S (2009) Practical declarative network management. In: Proceedings of the 1st ACM workshop on research on enterprise networking. ACM, pp 1–10

Lim H-T, Weckemann K, Herrscher D (2011) Performance study of an in-car switched ethernet network without prioritization. In: International workshop on communication technologies for vehicles, Springer, Berlin, Heidelberg

Loo BT, Condie T, Garofalakis M, Gay DE, Hellerstein JM, Maniatis P, Ramakrishnan R, Roscoe T, Stoica I (2009) Declarative networking. Commun. ACM 52(11):87–95

Meyer P, Steinbach T, Korf F, Schmidt TC (2013) Extending IEEE 802.1 AVB with time-triggered scheduling: a simulation study of the coexistence of synchronous and asynchronous traffic. InL Vehicular networking conference (VNC). IEEE, pp 47–54

Steinbach T, Lim HT, Korf F, Schmidt TC, Herrscher D, Wolisz A (2012) Tomorrow's in-car interconnect? A competitive evaluation of IEEE 802.1 AVB and Time-Triggered Ethernet (AS6802). In: Vehicular technology conference (VTC Fall). IEEE, pp 1–5

Steiner W (2010) An evaluation of SMT-based schedule synthesis for time-triggered multi-hop networks. In: 31st IEEE real-time systems symposium. IEEE

Thangamuthu S, Concer N, Cuijpers PJ, Lukkien JJ (2015) Analysis of ethernet-switch traffic shapers for in-vehicle networking applications. In: Proceedings of the 2015 design, automation & test in europe conference & exhibition. EDA Consortium, pp 55–60

Thiele D, Ernst R, Diemer J (2015) Formal worst-case timing analysis of Ethernet TSN's time-aware and peristaltic shapers. In: Vehicular networking conference (VNC). IEEE

Zhang L, Goswami D, Schneider R, Chakraborty S (2014). Task-and network-level schedule co-synthesis of Ethernet-based time-triggered systems. In: Design automation conference (ASP-DAC), 2014 19th Asia and South Pacific. IEEE, pp 119–124

How IOT Based Automated Driving Can Help Cities to Reduce Air Pollution

Ralf Willenbrock and Jörg Tischler

Abstract This paper looks at the relation between IOT (Internet of Things) assisted automated driving and the challenge of air pollution causing mobility access restrictions for European cities. After a brief discussion of the EC regulatory framework, the need for additional pollution data will be described suggesting to shift existing single-source air quality models to 2D and even 3D measurement platforms for improved air quality monitoring. Finally, the role of telecommunication and ICT industry for automated driving solutions will be elaborated, following by the description of pilot services contracted in the Horizon 2020 EU funded Autopilot project, currently under project execution in several European large-scale pilot sites.

Keywords IoT · Automated driving · Air quality · 5G · Traffic data · Horizon 2020

1 Introduction

When it comes to air pollution, many European cities are facing serious challenges to comply with existing environmental laws and policy regulations. NO_2 and PM measurements regularly exceed WHO thresholds and policy makers started to implement Clean Air master plans to reduce air pollution. Road traffic is reported to be responsible for almost 40% of emissions[1] and clean air measures vary from driving bans to low emission zones and city tolling. Giving the importance of mobility and public health for European citizens, many innovative concepts for low emission mobility are in the development phase, e.g. the EU funded Autopilot project with a focus on IOT based Automated Driving for electric vehicles.

[1] Bundesumweltamt (German Environmental Protection Agency): Nitrogen dioxide emissions. In: https://www.umweltbundesamt.de/daten/luft/luftschadstoff-emissionen-in-deutschland/stickstoffoxid-emissionen#textpart-1.

R. Willenbrock (✉) · J. Tischler
T-Systems International GmbH, Fasanenweg 5, 70771 Leinfelden-Echterdingen, Germany
e-mail: Ralf.Willenbrock@t-systems.com

J. Langheim (ed.), *Electronics Components and Systems for Automotive Applications*, Lecture Notes in Mobility,
https://doi.org/10.1007/978-3-030-14156-1_17

It is assumed that Automated Driving will support cities to give better urban access for commuters when looking for Park & Ride facilities as well as for alternative shuttle services and last mile logistics. For example, air pollution could be reduced by measures of offering combined tickets for commuters including free parking. In case of exceeding certain limits of air pollution, the public transport fee would then be reduced, even including Park & Ride access for car drivers. In 2017, the German Ministry of Transport and Digital Infrastructure (BMVI) launched a One-Billion-Euro Program to help German municipalities[2] to implement clean air master plans and cut down NO_2 emissions until 2020 to become compliant with EC air quality regulations again, thus avoiding any type of legal actions already threatened by the European Court of Justice.

2 Environmental Risks and Mobility Challenges for Urban Transport Systems

One of the key problems in cutting down traffic related NO_2 emissions lies in the complexity of finding strategic priorities for reduction measures. To derive strategic measures for this target, Newtonian Physics gives a mathematical approach by applying the energy equation for an object in motion. The equation for fuel consumption in units of [Liter/100 km] is written as

$$\Phi(v > 0) = \eta b_e \frac{\int_o^T (F_{acc} + F_{brake} + F_{roll} + F_{air} + F_G) v(1s) dt}{\int_o^{T'} v(1s) dt} \tag{1}$$

$$\Phi\left[\frac{\text{Liter}}{100\text{ km}}\right] = \Phi(v > 0) + \Phi(v = 0) \tag{2}$$

$$F_{acc} = m * \frac{dv}{dt}, \quad dv > 0 \tag{1a}$$

$$F_{brake} = \beta m * \frac{dv}{dt}, \quad dv < 0 \tag{1b}$$

$$F_{air} = \frac{\rho}{2} * A * c_w v^2 \tag{1c}$$

$$F_{roll} = mg\,\mu \tag{1d}$$

$$F_G = mg * \sin(\alpha) \tag{1e}$$

$\eta = engine\,efficiency\ [\%], b_c = fuel\,value\ \left[\frac{\text{MJ}}{\text{Liter}}\right]$
$m = total\,weight\ [\text{kg}], v = speed\ \left[\frac{\text{km}}{\text{h}}\right], \beta = propulsion$

[2]BMVI—Bundesministerium für Verkehr und digitale Infrastruktur (Federal Ministry of Transport and Digital Infrastructure): Clean air 2017–2020 initiative. In: https://www.bmvi.de/SharedDocs/DE/Artikel/DG/sofortprogramm-saubere-luft-2017-2020.html.

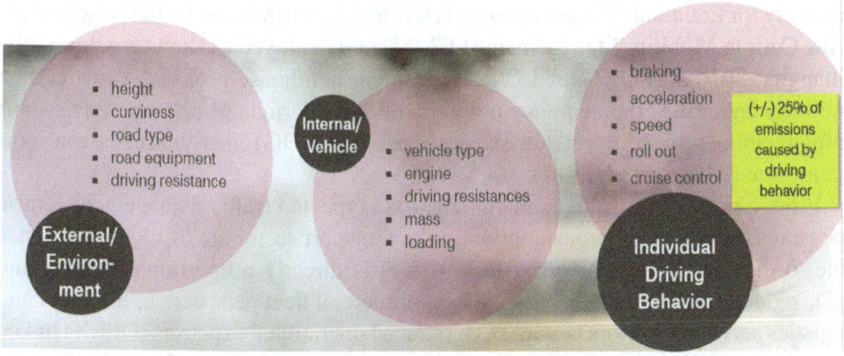

Fig. 1 Influence factors for NO_2 emissions in vehicle based road traffic

$\mu = friction\,coefficient,\, g = acceleration\,of\,gravity$
$\rho = air\,density,\, A = Cross\,sectional\,area,\, c_w = drag\,coefficient$
$\alpha = gradient,\, T = driving\,time$
$T' = Travel\,time\,to\,reach\,the\,reference\,distance,\,usually$ 100 km.

For further details see reference (Reif et al. 2011). Despite of the numerous input parameters in Eqs. (1) and (2), there are 3 important categories to mention:

(a) time-dependent parameters per trip (speed, acceleration, travel and driving time, gradient, standstill);
(b) vehicle depending parameters (weight, cross sectional area, drag and friction coefficient, engine efficiency and fuel value);
(c) physical constants (acceleration of gravity, air density).

This leads to the possibility of measuring the energy consumption per vehicle trip by collecting GPS speed profiles per second with the help of a simple smartphone. By comparing these speed profiles with well-known and standardized driving cycles, e.g. the European Driving Cycle, and reference vehicle configurations, the deviations of driving resistances as stated in (1a)–(1e) can be analyzed in [%]. This helps to better understand the cause for low and high fuel consumption occurring during a trip, i.e. in complex urban road conditions. As shown in Fig. 1, such an analysis per trip enables fleet operators to cut down fuel costs and emissions by having a complete picture of all influence factors for emissions at the same time, which are

1. Analysis of Driving Behavior leading to better "predictive driving and braking",
2. Analysis of Road Types leading to better "traffic-optimal speed selection" and "routing plans",
3. Analysis of Traffic Conditions leading to "congestions minimized tour plans".

In the years 2008–2012, T-Systems developed and implemented a system to operate such strategic fuel reduction measures as Cloud based service for Logistics Service Operators, which was funded by Deutsche Telekom AG and the German Federal

Ministry for Economic Cooperation and Development (BMZ) under the project name Low Carbon Mobility Management (LCMM) and deployed as showcase during the Shanghai 2010 EXPO. Thus, LCMM became a lighthouse project for low carbon mobility under the climate policy umbrella of Sino-German Ministries of Transport implementing projects to cut down Greenhouse Gas (GHG) emissions in the transport sector, see (Nocera et al. 2017).

From a policy perspective, all vehicle, road type and traffic management centric measures to reduce fuel consumption in the transport sector are of high importance due to the fact that every saved liter of fuel is directly proportional to CO_2 and NO_x reductions. On the other side from commercial fleet perspective, transport and logistics service operators benefit from reduced operational expenses (OPEX) linked to fuel costs. After the successful pilot implementation with two Logistics Service Providers in Beijing and Jiangsu (China), LCMM was rolled out in Europe in several logistics R&D projects[3] and is now accepted for 2019 ETSI standardization in the Working Group Nomadic Devices, which is part of the Horizon 2020 project AEOLIX (see www.aeolix.eu).

Besides efforts towards standardization of carbon footprint monitoring by the use of nomadic devices, the first Chinese Logistics Service Providers (LSP) decided to operate LCMM directly within their fleet operation centers. Since 2015, more than 200,000 h and 10 Million km of fleet operations were reported by GPS smartphones leading to reduction measures of eco-drive training and optimized tour planning. This was achieved by weekly scoring reports and applied Big Data analysis made available directly to the operating staff which had regular dialog with the drivers delivering goods. After implementing LCMM for the first Chinese LSP, an average of 16% fuel and carbon savings per ton of payload was found in the financial systems, thus supporting the sustainability and efficiency targets of the LSP user.

As LCMM will be submitted within Aeolix for ETSI standardization, the procedure of applying energy Eqs. (1) and (2) as KPI for impact evaluation can be considered helpful for upcoming new mobility projects such as Autopilot, an IOT based Automated Driving project within the H2020 R&D framework.

3 The Holistic Approach: From Single to Multiple Source Analysis of Air Pollution in Urban Transport Mapping Complex Cause and Effect Phenomena

With regards to air pollution, data collection and KPI impact evaluation, e.g. for Sustainable Urban Mobility Plans (SUMP), play a crucial role to find out the effectiveness of actions taken by a municipal government in the context of cutting down NO_2 emissions according to the environmental policy regulations for Public Health

[3]Further LCMM project reference details in: https://business.esa.int/projects/eu-wide-mobility-safety-efficiency-management-logistics or http://cogistics.eu/ or https://energypedia.info/wiki/Emission_Data_Monitoring_Technology.

as agreed by all EU member states. The key challenge remains the reliability of pollution measurements as the number of measurement stations is limited and extremely sensitive to the space-time model of detection.

Usually, legal actions by the European Court of Justice are based on measurement stations authorized by Environmental Agencies in the member state which signed the 2008 EC Environmental regulatory framework. Violations lead to legal investigations and in worst case to fines based on a corresponding court decision. Nevertheless, the baseline for all legal and political decisions are single source measurements located in very limited space and time under control and authorization by Environmental Agencies.

Given the complexity of phenomena influencing the overall emissions in an urban environment a fine-tuned control loop (see Fig. 2) to cut down emissions is required. In such a control loop the first Phases 1 and 2 include the realistic planning and definition of reduction targets. For this purpose, existing emission modelling is recommended to be expanded by road network data delivered by Public Authorities, mobile speed data collection from cellular telecommunication networks and Floating Cars from urban logistics fleets and Public Transport operators. When creating the appropriate data fusion algorithms, 1-dimensional single-source modelling can shift to 2-dimensional multi-source modelling. For such an approach Automotive Industry has a lot of useful mobility data to share with Public Authorities and 20 years of project experiences along TMC standardization for Dynamic Navigation Systems is a good examples of innovative service evolution, see (Willenbrock et al. 2006; Breitenberger et al. 2004). Additionally, first experiences from using Earth Observation satellite data for the integration of ground-based air quality measurements by Copernicus atmosphere data merged via data fusion towards 3-dimensional air quality information started and can be considered as the next exciting innovation path for air quality services in the near future.[4]

Compared to the single-source approach, such an innovation pathway towards 3-dimensional analysis of air quality offers long-term control of achieving the political emission reduction targets which are inside the European Union very strict and therefore difficult to achieve. Nevertheless, one economic objective of the EU-wide strict regulations was to foster industrial product innovation bringing the necessity of close cooperation between automotive and telecommunication industry with municipal traffic and environmental authorities. A schematic illustration of the holistic approach is shown in Fig. 3.

In Fig. 4, some typical clean air initiatives are listed ranging from speed reductions, access restrictions, combined public transport ticketing strategies or low emission concepts for urban goods delivery. Each of these measures must be controlled by municipal decision makers in an authorized manner to find out long-term effectiveness or political need for changing or adjusting a clean air action plan. Therefore, the shift from single to multiple source analysis is needed, a process which took place in weather forecast by establishing the science of synoptic meteorology with

[4]Further details. https://www.dlr.de/dlr/desktopdefault.aspx/tabid-10261/371_read-2869/#/gallery/5158.

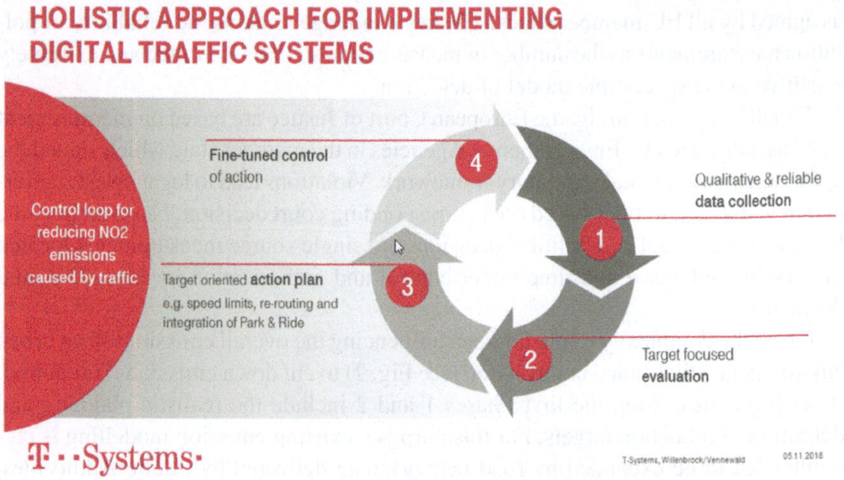

Fig. 2 Control loop for reducing NO_2 emissions caused by road traffic

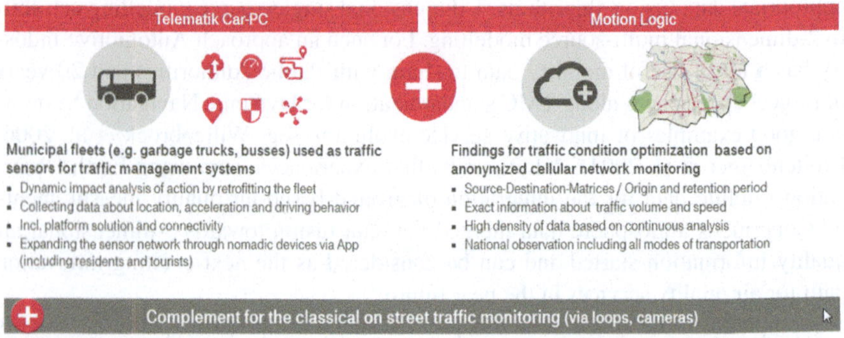

Fig. 3 Extended data collection strategies based on multi-source integration of fixed and mobile data from floating cars with telematics devices as well as anonymized cellular speed profiles from telecommunication networks

the introduction of 3-dimensional weather modelling through synoptic evaluation of merged stationary and satellite data sources. For the similar complex task of NO_2 reduction, the shift to 3-dimensional data analysis of air quality is promising as it might lead to the same quality in forecast models for clean air action plans. This is illustrated in Fig. 4, where one can see the different building blocks for such a 3-dimensional data analysis and control loop methodology suggested for European municipalities.

Before progress towards synoptic clean air modeling can take place, new mobility data sharing platforms have to be built in close cooperation with data driven business models. This is the same for Automated Driving when being extended to larger areas of road segments in cities. Here, IOT sensor data exchanged with high performance cloud platforms is assumed to play a key role for IOT based Auto-

Fig. 4 From 1-dimensional to 3-dimensional air quality measurements, including NO_2

mated Driving. In 2016, the European Commission funded five Large-Scale Pilots (LSPs) on the Internet of Things.[5] The AUTOPILOT project was selected as Pilot 5 for autonomous vehicles in a connected environment. The EU reviewers wanted to showcase Automated Driving in complex road conditions with the support of IOT and the perspective of attractive mobility use cases with positive impact for clean initiatives. Mobility data sharing is one of these use cases.

[5]Further project background in: http://autopilot-project.eu/.

4 The Role of Telecommunication and ICT Within the Autopilot Project: Balancing Mobility Needs with Access Restrictions for Citizens Caused by Clean Air Initiatives

After describing the need of multiple source analysis for municipal decision makers when implementing clean air master plans including mobile data from vehicles and their speed profiles along the road network, the role of telecommunication and ICT industry for IOT assisted Automated Driving services needs to be introduced. The Low Carbon Mobility service in Fig. 1 is based on mobile communication and needs real-time data provisioning (connectivity). Compared to this service the control loop from Fig. 2, is running on a political decision base and is independent of communication technology when using stationary measurement stations alone. Nevertheless, given the importance of mobility access restrictions for commuters and urban citizens the reliability of data collection and the effectiveness of the Clean Air policy measures shows that the ad hoc availability of data helps municipalities to increase transparency and to better convince citizens of taken measures. With the upcoming 5G technology and the standardization behind it, complex data exchange between sensors, data centers, road side units and vehicles can be managed and guaranteed for users and mobility services.

As shown in Fig. 5, the road side infrastructure (traffic lights, VMS, etc.) can be connected to the vehicle either by I.T.S.-G5 or LTE-V, complemented in the next years by cellular (C-) I.T.S., just depending on the different local implementations. Additionally, mobile communication data, e.g. from Motionlogic, supports this set-up for both, LCMM and Automated Driving.

Within the Horizon 2020 Autopilot project,[6] all active pilot sites will exchange data both from road side units or via cellular networks. Mainly, such hybrid communication approach is used because Autopilot wants to showcase services and business cases in which innovative mobility services are operated in complex road conditions of several kilometers length but is not limited to service such as automated valet parking in limited geographical space of some hundreds of meters, where only road-side units with point-to-point vehicle-to-infrastructure communication (V2I) are available, e.g. car parks. Figure 5 also illustrates that from the telecommunication network side, mobile edge computing (5G MEC) is under investigation for car-to-car communication for low latency reasons.

It is well-known that automated driving will reduce carbon emissions, but this qualitative impact is difficult to quantify in certified KPI's needed for SUMPs. And this is the link to standardized carbon footprint monitoring services such as LCMM used in Aeolix. With the help of standardized Big Data analysis along the energy Eqs. (1) and (2), the effectiveness of IOT assisted Automated Driving can be evaluated

[6]See Footnote 5.

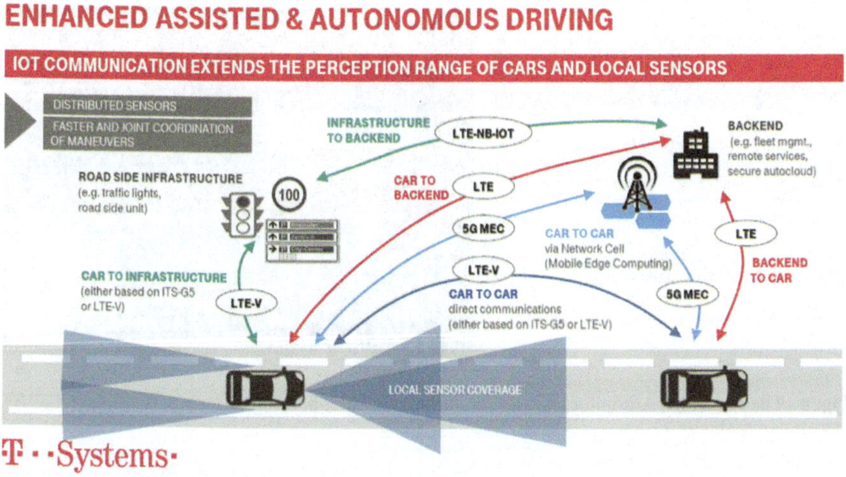

Fig. 5 Schematic overview of the different data exchange objects relevant for IOT based automated driving service deployment

and the measurable impact analysis allows mobility service providers to quantify the amount of NO_2 and GHG savings achieved by the deployment of innovative new services.

Within the Autopilot project, environmental impact is of special interest for all mobility services under development.[7] Figure 6 shows this in exemplary manner. The Versailles pilot area is illustrated on the map showings that tourists can book AD vehicles and drive either by themselves (inside the city of Versailles) or in an automated manner (inside the park of the castle). After dropping the vehicles off somewhere inside the city, a manned lead vehicle will collect the AD vehicles in a platoon and bring them back to the parking space of the rental car service provider. This way, the mobility service provision is directly linked to tourism, which is foreseen to massive growth rates shown in Fig. 6. It is assumed that the number of tourists visiting Europe could increase from 609 million visitors in 2015 up to 1.8 billion visitors in 2030, which would be an increase of more than 300%. As tourists generally are not familiar with the road network and local traffic conditions, Automated Valet Parking or City chauffeur rental car services are considered as attractive service offering for a touristic venue in any European country.

The idea of bridging such innovative Automated Driving services with holistic control loops for NO_2 reduction targets, including automotive and telecom combined mobile data collection for environment-sensitive traffic management tasks presented in Chap. 3, seems to open the door for new data markets. As the quality of urban transport modes and metropolitan air quality decrease or increase in a synchronized manner, the holistic approach and research for Automation seems to be natural way of innovation, municipalities are looking for. Here, the Autopilot project is a good

[7]In: http://autopilot-project.eu/autopilot-library/publications-and-presentations/.

Fig. 6 The versailles Autopilot test site for automated valet parking

example of bringing together industry, research and public-sector stakeholders into one project and can be considered as lighthouse project for innovative mobility services.

5 Summary

Nowadays, mobility access restrictions for low emission zones are widely spread among European cities, mostly initiated to comply with the strict air quality regulations of the European Commission. Nevertheless, the effectiveness of all air quality master plans is easily questioned by the general public as long as only single source air quality detection is in practice, because this makes air quality measurements in complex urban environments quite incorrect and sensitive to statistical and rather local phenomena. Therefore, additional mobile and telematics data sources from vehicle fleets and mobile communication networks are recommended to shift 1D single source detection towards 2D or even 3D air quality maps. In this paper, it was elaborated how such an approach could help clean air initiatives similar to the quality improvement which took place in weather forecast by moving towards synoptic meteorological models with the help of Satellite data sources. Once the data sources are more reliable, European municipalities will have to implement innovative mobility services compliant with improved air quality detection. This can include intermodal shift and extended Park-and-Ride offerings, in especially for commuters. In the paper it was shown that IOT assisted Automated Driving, is an excellent complementary technology for Cities looking for low emission mobility services as it allows to establish flexible car sharing or rental car offerings. In the Horizon 2020 project Autopilot, these offerings are currently under investigation in several large-scale pilot sites and linked to tourists venues, with French City of Versailles being one of them.

Literature

Breitenberger S et al (2004) Extended floating car data—Potenziale für die Verkehrsinformation und notwendige Durchdringungsraten, BMW Group Research. Straßenverkehrstechnik

https://ec.europa.eu/germany/news/20180517-luftverschmutzung-klage_de

Nocera, Willenbrock et al (2017) Reducing fuel consumption and carbon emissions through eco-drive training. Transp Res Part F 46:96–110

Reif et al (2011) Robert Bosch GmbH, automotive aftermarket, 27th edn. In: Bosch Kraftfahrtechnischen Handbuch

Willenbrock R, Steinert F, Graze K, Schönewolf W (2006) EVENT-ONLINE—A service concept or large scale events based on FCD technology. In: Advanced microsystems for automotive applications, pp 41–44. https://link.springer.com/chapter/10.1007%2F3-540-33410-6_4

Holistic HMI Architecture for Adaptive and Predictive Car Interiors

Frédéric Fonsalas

Abstract With the fast-ongoing evolution towards C.A.R.E. vehicles (Connected, Automated, Ride-Sharing, Electrified) interior arrangements are deeply transformed. Both seats and surfaces as part of the interior are becoming populated with actuators and sensors in order to make them truly adaptive and predictive to the various situations encountered while driving. This evolution leads to the perception of the full interior as an extensive "Human Machine Interaction (HMI)" system providing customized comfort, safety, smart and enjoyable life-on-board in all situations. The main benefit for the car users is utilizing the driving time for productive activities in continuity with their occupations outside of the car. This paper explains the end-to-end architecture ("Cockpit Intelligence Platform"—CIP) views of Faurecia for such holistic interior HMI solutions. Faurecia is a worldwide leader in car interiors which has created an extensive partnership network in the last years in order to meet these new challenges. Faurecia demonstrated its concrete solutions and innovations through prototypes and demonstrators at the Frankfurt Motor Show 2017, Las Vegas CES 2018 and recently at the Paris Motor Show, where the company attracted a considerable attention from the automotive community.

Keywords Connected · Automated · Ride-Sharing · Electrified · Electric Vehicle · Cockpit · Cloud platform

1 Introduction

It is obvious that humankind is developing at an exponential rate. One example: the number of human beings populating the earth. Each generation includes as many individuals as the sum of all those who lived before it, since millennia. This is not a fact of nature, as other species do not undergo the same curve, but it is clearly

F. Fonsalas (✉)
Director Strategy Parrot-Faurecia Automotive, Faurecia, 174 quai de Jemmapes, 75010 Paris, France
e-mail: frederic.fonsalas-ext@faurecia.com

J. Langheim (ed.), *Electronics Components and Systems for Automotive Applications*, Lecture Notes in Mobility,
https://doi.org/10.1007/978-3-030-14156-1_18

the result of humans' capability to understand, develop, and hopefully manage their environment and its resources.

In the last 100 years, significant developments in the world, starting with the transportation revolution (the invention of combustion engines), followed by the Information revolution (stemming from fundamental discoveries in physics such as the transistor and the LASER), have been powerfully fueling extremely large and radical evolutions not only of the products at our disposal, but more fundamentally of our lifestyles.

We have come now to the point where these two revolutions (Transportation and Information) are merging and giving birth to probably even further reaching transformations with countless consequences. New types of vehicles such as semi-autonomous or fully autonomous cars are rapidly brought on the market. In parallel, virtuality increases as well, making it possible to realize remotely what was previously requiring a physical move, e.g.: teleconferencing, online shopping, working, studying, entertainment, etc.

Some researchers depict the evolution phase we are in as the VUCA, which means a Versatile, Uncertain, Complex and Ambiguous world. Automotive products are spectacularly impacted by this move. As a consequence of the general VUCA context, they become radically (Fig. 1):

- **Connected**, meaning that all data and information we use in other contexts are made available in the car as well;
- **Autonomous**, meaning that the car becomes a robot, relieving the driver from the task of driving under increasingly stricter conditions (jams, regulations, controls…)
- **Ride-Sharing**, meaning that cars are seen less as possessed goods, more as a commodity that can be obtained when needed, and payed use-by-use;
- **Electrified**, pointing a radical change in the propulsion principle itself, evidently fueled by environmental concerns. It has enormous consequences such as a drastic simplification of the motoring technology and the possibility of far simpler and better car interior designs.

A particular consequence of these changes is the proliferation of new actors in the field of mobility, following the Tesla model. Lots of them embark on the electrical

Fig. 1 C.A.R.E. vehicles and their benefits

propulsion mode (Tesla, Rivian, Lucid…), some on the pure mobility service (such as Uber or Didi Chuxing), and others come from the "virtual world" (Google, Apple or Baidu).

This pushes traditional players to rapidly and deeply reconsider their traditional technology, competency basis and their business models. For example, the German automotive industry that traditionally focused on horse-powered combustion engines, is now reengineering itself into breakthrough electric propulsion engines.

2 Evolution of Automobiles Architectures

At the level of car architectures and technology, these changes have far-reaching consequences.

Overall, automotive products will gradually be limited to the assembly of three "macro" components:

- **The chassis** (including engine), populated with sensors and intelligent software enabling them to direct themselves in an autonomous manner, even in complex situations;
- **The cockpit** (meaning the complete interior of the car), increasingly becoming a "moving lounge";
- **The cloud**, establishing the link with the external world and enabling a full continuity of activities in or out of the car;

Parrot-Faurecia view is that this macro-partitioning should be reflected in the long run by the electronic architecture through four technical platforms (Fig. 2):

- A cloud intelligence and service platform
- A connectivity platform

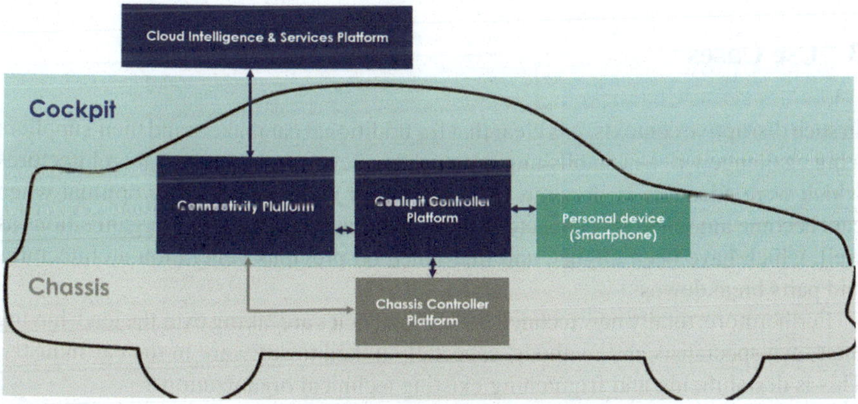

Fig. 2 Parrot-Faurecia view on the macro-partitioning of automotive products

- A cockpit controller platform
- A chassis controller platform.

Each of the three macro components will follow specific logical development routes:

- The chassis is becoming a robot; Artificial Intelligence and highly sophisticated real-time short-loop control and supervision algorithms are clearly keys to this domain. The chassis will need to interact with its environment and therefore will have the ability to communicate. The 5G standard currently in deployment represents a much larger step than previous 2G/3G/4G mainly because it includes standards for "Internet-of-Things" (IoT) objects in general and Vehicle-to-Things" (VtoX) protocols and capabilities. It is also clear that chassis will undergo stricter approbation procedures, drastically narrowing-down the number of vendors of their key modules, particularly, the Autonomous Drive Controller.
- The cockpit is evolving along two fundamental lines: it becomes "another living room or office", and it is the element of diversification and identity of the car. Comfort, communication devices like large displays, outstanding sound and lighting systems, full continuity with pre and post-car activities through connectivity, aesthetics & materials quality are the key directions for connected/autonomous cars interiors. Shared vehicles will introduce the additional requirement of personalization: different people using the same car in specific time slots and situations, or several persons involved in different activities with specific preferences sharing the same trip. This creates the "bubble" topic as one of the major technical challenges for example, in the audio, entertainment and thermal comfort domains.
- Finally, cloud servers are obviously key-enablers of car-customization in different contexts: the continuity of activities for individual car occupants, fleet management for professionals, rental car companies, cities, geographically localized ecosystems (Google, Yandex, Baidu...). But progressively, through big data, cloud servers will also enable a higher level of service than what an "isolated vehicle" could deliver.

3 Use Cases

In such disruptive contexts, it is clear that for traditional car makers and their suppliers a big challenge will be establishing a legacy. Legacy refers to technical architectures which were adequate in previous generations but no longer valid or optimal when cars become autonomous connected robots. This can be the case for organizations as well, which have been aligned and fine-tuned on previous generation architectures and parts breakdowns.

Furthermore, totally new technology competencies are taking over the lead, having their own specialists and methods, especially linked to software in the car industry. This is destabilizing and frightening existing technical organizations.

In this context, the temptation to proceed "incrementally" for evolving architectures and organizations is very understandable. Unfortunately, it is associated with a major risk of developing obsolete solutions with huge efforts and investments.

As an example, I was recently in a taxi somewhere in northern Europe. The car, produced by a recognized leading innovator in electric vehicles, had a huge display and a nice map of the country on its dashboard. Still the taxi driver was using his smartphone, probably 20 times smaller, less comfortable to use and 100 times cheaper than his car, to navigate.

As painful as it could be, it is essential to remember that architectures and organizations are solutions for offering relevant use cases to customers, not self-justifying realities. And, as a matter of fact, with the transition from cars to C.A.R.E vehicles, totally new use cases are emerging, e.g.:

- In-car office;
- In-car cinema;
- Individual audio bubbles;
- Personalized climate comfort;
- Versatile cockpits, where seats for example can be rearranged from autonomous to driving modes;
- Individual customization;
- Occupants health monitoring.

This is the reason why Faurecia has followed a logic starting with use-cases and then focusing on cockpit architectures for reengineering its capabilities and organization, deciding its alliances and acquisitions and creating user-friendly next-generations cockpits.

At Faurecia everything starts from a system-level "Cockpit-of-the-Future" activity which is imagining, exploring, representing and testing the new use cases of years 2020–2030, driven by the C.A.R.E megatrends.

4 Technical Challenges

Looking into such use cases, Parrot-Faurecia has tackled several key technical challenges:

- The mere assembly of elementary functional "components" no longer does the trick. Two main factors are at play:
  - On the one hand, use cases cannot be handled by single isolated technical components anymore (e.g. a car radio, a seat, a button…). Managing use cases involves a complete chain going all the way down from the user interface (sensors, actuators), through control and intelligence modules, connectivity channels (4G LTE, 5G, V2X…) up to cloud-based software and services, thus creating end-to-end services.

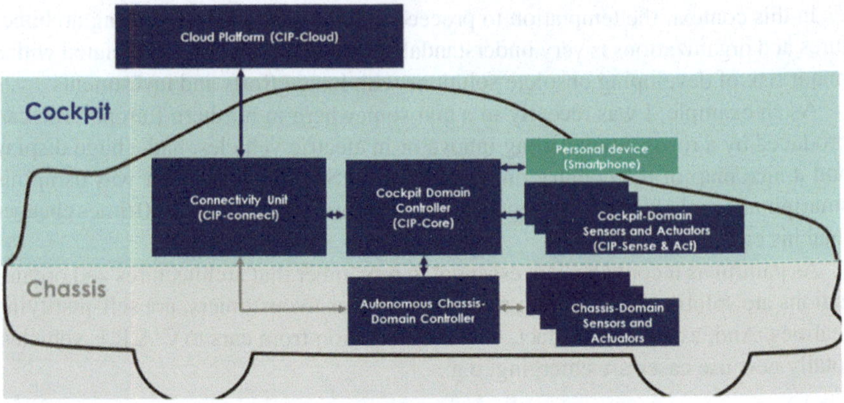

Fig. 3 Parrot-Faurecia frame-of-reference for next-generation vehicle electronic architectures

– On the other hand, individual sophisticated services alongside one another (for audio, lighting, infotainment, health, thermal, safety, etc.) will never make for a consistent, unified and useable system. Cockpits of the future also imply proper holistic, transversal and user-friendly Human Machine Interface solutions.

- New use cases require traditional "static" components to "come to life". For instance, seats can now provide massage, heating and cooling, or move every which way. Door trim components are able to display images, generate light, deliver haptic feedback, become tactile and vibrate to generate high-quality audio and music. In a nutshell, all traditional mechanical components of car cockpits are becoming smart, adaptive, and even predictive.
- Software cannot be a "nice-to-have" element added as an afterthought on top of mechanical- and trim-based cockpits. Software is actually the lifeblood of next-generation vehicle interiors.
- Service continuity with the outside world is no longer optional. As driving is losing relevance, vehicle occupants get the opportunity to earn back time. Being in a vehicle is no longer a daily hassle. It is a new space, an extension of their living room or their office where they can relax, be entertained or productive—hence the need for relevant and up-to-date infotainment and content.
- Unprecedented technical issues arise as a result: running open ecosystems (e.g. Google, Yandex or Baidu) along with highly secure software modules on the same electronic platforms creates a whole range of cybersecurity concerns.

The resulting macro electronic architecture of vehicles is Parrot-Faurecia frame of reference (called the "Cockpit Intelligence Platform—CIP") and involves four domains (Fig. 3).

1. a **cockpit domain controller ("CIP-Core")**, associated with **cockpit domain sensors and actuators**, short-loop ECUs (**"CIP Sense & Act"**);

2. an **(Autonomous) chassis domain controller**, associated with **chassis-domain sensors and actuators**, short-loop ECUs (*Not in Faurecia's perimeter*);
3. a **connectivity unit ("CIP-Connect")**;
4. a **cloud platform ("CIP-Cloud")**.

As Faurecia is focusing on car interiors, we will now dive deeper on items 1, 3 and 4.

The cockpit domain controller (CIP-Core) is an evolution and a fusion of several traditional cockpit ECUs:

- The **instrument cluster**: its main purpose is to provide the driver with any information relevant to driving, to the vehicle status and/or. Over the last few years, this ECU has shifted from mechanical and analog user interfaces (pointers and gauges) to multi-purpose freely programmable digital displays
- The **In-Vehicle Infotainment (IVI) unit**, stemming from the car radio and enhanced over the years by multimedia, navigation and connectivity with nomadic devices (such as smartphones)
- The **HVAC control unit**: what was originally a purely mechanical heating system has turned into a smart and programmable comfort interface over time
- **Specific body functions** like seats position adjustments, health and wellness controls.

Given the development of programmable technology solutions (essentially large programmable touchscreens), the automotive industry is increasingly taking the focus off driving itself to redirect it to other activities linked to entertainment, communication and information. The emergence of autonomous vehicles will completely relief the driver from the hassle of driving in certain situations (levels 3 "eyes off" and 4 "mind off" of driving automation). Depending on the level of automation along the trip, the vehicle will thereby seamlessly shift from a traditional driving environment to an actual "office on wheels" tailored to the user preferences and habits.

The underlying technical solutions require the combined development of several innovation streams:

- Large displays integrate once dissociated functions into one surface (IVI + HVAC, and even the instrument Cluster);
- Powerful "Systems-On-Chips" (SOCs) run multiple functional domains on one CPU/GPU;
- Software partitioning and virtualization enable to run multiple operating systems (e.g. Android and ASIL-certified real-time OS) independently on the same platform;
- Over-The-Air (OTA) updates go increasingly deeper in the software layers (middleware, firmware) even within the lifetime of a vehicle;
- Cybersecurity is paramount at all levels of the chain: software modules, networks, factories, cloud, etc.

The connectivity unit (CIP-Connect) is the key node where data transfer, control and distribution occur, from the vehicle to the outside world and vice versa. It is mainly made up of two blocks:

- A **reception unit** to demodulate or modulate data from several standards. Traditional broadcast standards such as analog and digital radio will not disappear overnight. However, 2-way channels are increasingly gaining ground. 4G LTE and soon 5G cover a wider scope than mere telephony: Internet access, traffic information, MoD or even VoD—richer and better content for a complete infotainment suite. This unit will also include short distance wireless technology standards such as Wi-Fi and Bluetooth;
- A **data analysis unit** to interpret data and route it to the proper endpoint. This connectivity gateway plays a fundamental role in safety and security as it stands where in-car networks and external data networks meet.

The cloud platform (CIP-Cloud) provides the car with up-to-date applications, data, and algorithms:

- As of now, Android-based IVI systems already use cloud servers to offer seamless customization tailored to the preferences of the end-users and car makers:
  - **Geographically relevant** content and ecosystems (Google in Europe, North and South America, Yandex in Russia, Alibaba and Baidu in China…)
  - **Specific re-skinning and app store** (for shared vehicles platforms across brands for example)
  - **Scalable** (integrating new popular apps that were not available at production start)
- In the coming years, thanks to the collection of end-users preferences and habits combined with cloud-based big data analysis and artificial intelligence, car functions will display new and smarter behaviors delivering a better interior user experience.

5 New Business Challenges and Models

The extremely deep transformations covered in this paper naturally have consequences on the business models and the role of the different actors in the value chain. We do not claim to draw up an exhaustive list of the different possible configurations, as they will result from a complex mix of:

- Actual technology choices
- Legacy situations for traditional OEMs
- Greenfield approaches of new entrants
- Emergence of new technical actors
- Expansion of existing Tier-1s in new territories.

Faced with this complex and uncertain future, Parrot-Faurecia strategy focuses on adequate answers to two overwhelming challenges:

1. Creating "agnostic" banks of software service modules that can be integrated in a controlled, cost-effective and rapid manner across different specific vehicle software environments
2. Developing system modeling tools bridging the gap between the description of user experiences and holistic systems able to consistently orchestrate said experiences.

It should be highlighted that these new business models will be executed along two holistic axes: with services being the "vertical" axis and the overall system, including the Human-Machine interfaces the "horizontal" one.

The first axis is **the end-to-end service axis**, represented in red on Fig. 4. As explained above, emerging in-car services require a "vertical chain" of technical modules running all the way from the cloud down to the end-user in the vehicle. It is essential to have a dual expertise in physical components (actuators, sensors, displays, loudspeakers, etc.) and virtual components (software short-loop algorithms, AI-based control algorithms, big data, etc.) to be able to deliver specific services actuating physical vehicle parts (seats, lighting or audio). Combining mechanical and electronic know-how is a key underlying trend explaining recent M&A moves among major automotive Tier-1s.

The second axis is the overall system including the **Human-Machine Interface dimension**. All "vertical services" need to be combined together. The sheer number of new services and the speed at which they are developed create unprecedented challenges when it comes to the usability and appeal of global vehicle systems such as:

- **Fast and reliable combination of services** into a complete car scenario, enabled by Service-Oriented Architectures in particular
- Creating and assessing **user-friendly and attractive interfaces** to manage the underlying service complexity

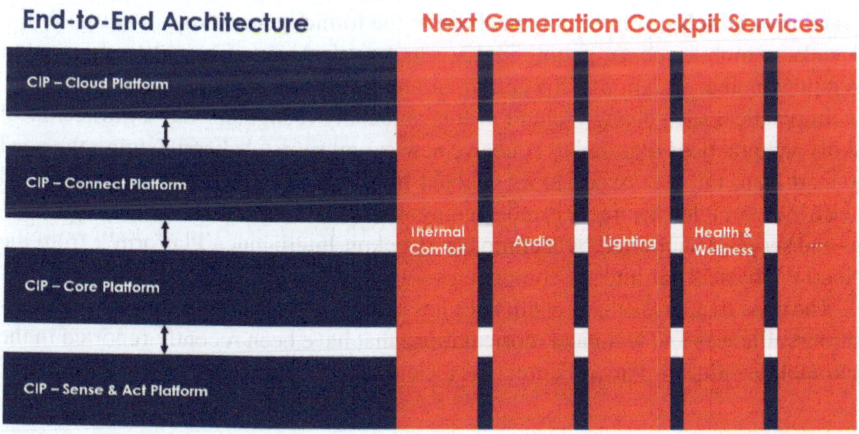

Fig. 4 "Vertical" end-to-end service axis

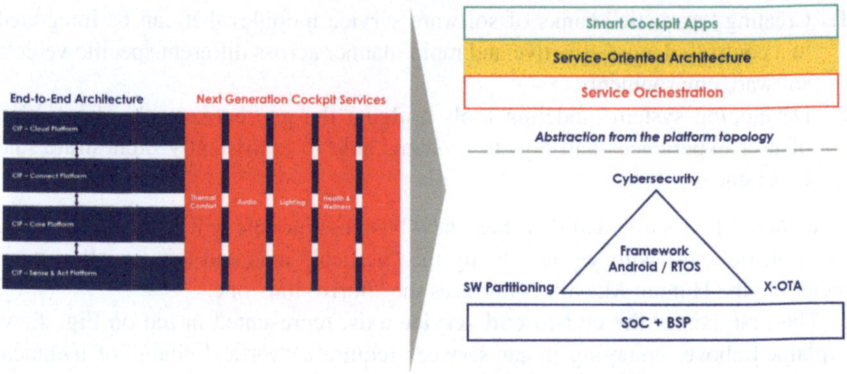

Fig. 5 "Vertical" services implementation on the "horizontal" topology & HMI axis

- **Authoring**: as programmable displays are taking up increasingly larger interior surfaces, graphic design and HMI logic will end up becoming core elements of the car makers brand DNA
- **Scalability**: as new services will pop up all along the decade-long lifecycle of most vehicles, HMI will have to keep up with those evolutions.

Figure 5 is showing how the "vertical" services are implemented onto the "horizontal" platform structure.

6 Summary

Over the last decades, technical breakthroughs in the automotive and communication industries have converged to such a degree that it is now possible to come up with brand new vehicles, similar only in shape to the former generations of automobiles. Thanks to mobile robots, getting into a vehicle is no longer focused on driving but is rather another situation in the continuum of daily activities.

In practice, such changes entail to fundamentally reengineer the set of technical skills and practices required to conceive new car interiors. Acknowledging the need to start from the user experiences enabled by autonomous, connected, shared and electrified vehicles, Faurecia is redesigning its engineering capabilities to offer end-to-end services supported by a complete "Cockpit Intelligence Platform": from the cloud to physical car interior components—its traditional know-how and business.

The need to address those challenges has also dictated a fair number of the acquisitions, alliances and technical reorientations that have been recently reported in the automotive industry. Parrot-Faurecia is a clear first-mover going in this direction.

Acknowledgements I would like to express my gratitude to Robin Denis, Product Marketing and Benchmarking Manager, and Melania van Laack, Digital Marketing Engineer, for their helpful suggestions and advice that contributed to this paper.

Part V
Connected Car, Privacy and Security

New Mobility Services and How They Will Be Affected by the Connectivity

El Khamis Kadiri and Antonio Eduardo Fernandez Barciela

Abstract Today, the automotive sector is under a huge transformation in the use cases and services due to two main vectors: automated driving and connectivity. Cars enter into the world of the Internet of Things to offer new experiences. The connected car has become an enhancer for some already existing use cases, an enabler for use cases that today have technical limitations and an opportunity for others where the automotive sector was not participating. There are huge expectations on 5G to bring up new opportunities thanks to the vertical approach taken this time, completely different from the previous mobile generation networks. The automotive is considered as one of the key verticals. The new paradigm for the Intelligent & Connected Vehicle is based on:

- Intelligent comes from being able to understand the needs of the driver and passengers thanks to a personal assistant which will be mainly in the cloud.
- Intelligent considered as automated driving, meaning that the vehicle will be able to take the control under different circumstances according to the SAE levels.
- Connected indicating the capability of getting information from outside the vehicle to complete the environment perception and to enhance some of the services provided to drivers and passengers.

Some of the key technology enablers, which can modify the use cases and the business models, will be exposed.

Keywords ICV (Intelligent and Connected Vehicle) · 5G · C-ITS · V2X

E. K. Kadiri (✉) · A. E. Fernandez Barciela
Groupe PSA, Research Innovation & Advanced Technologies Division, Connected Vehicle & EE Systems, Vélizy-Villacoublay, France
e-mail: elkhamis.kadiri@mpsa.com

A. E. Fernandez Barciela
e-mail: antonioeduardo.fernandez@mpsa.com

J. Langheim (ed.), *Electronics Components and Systems for Automotive Applications*, Lecture Notes in Mobility,
https://doi.org/10.1007/978-3-030-14156-1_19

1 Introduction

Connectivity is one of the major trends shaping the new automotive and mobility era, included in a key item list, namely: sharing mobility, emissions, autonomy, digitization, market divergence and customer behavior. From April 2018, the European regulation imposes all new cars to be connected, at least to provide the e-call service. Connectivity has important implications in the mobility services and particularly in the automated driving roadmap. The automotive industry is convinced of the need of the connectivity to complete de capabilities of the onboard sensor for high automated driving levels (from level 3 on). First, to provide information for completing the environment perception beyond the sensor line of sight limits. Second, to help in the decision-making, by providing network decision and control from outside the vehicle.

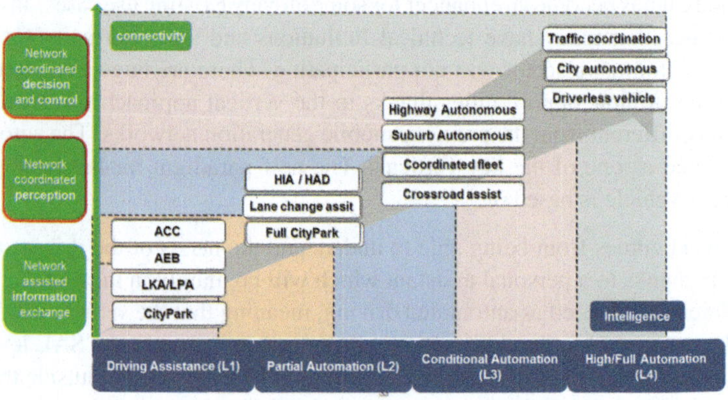

More precisely, the arrival of 5G brings new opportunities for mobility use cases which can be divided into 3 main groups:

- Existing services will be boosted and/or enhanced thanks to the connectivity performances improvement. Remote diagnoses, predictive maintenance and over the air updates use cases fall into this category.
- Regulated V2X use cases. In Europe, the European Commission is preparing a legal framework to define a regulation for day 1 and day 1, 5 use cases, defining the interaction of the connected vehicle with the environment: other vehicles, pedestrians and infrastructure. In China, the ministry of Transport has defined a whitebook with 72 different V2X use cases which will be promoted from 2020. The Chinese administration will be involved in this action by an important connectivity infrastructure deployment plan. The city of Wuxi, can be seen as a first example of this approach.
- Automated driving features for high automated levels will be more demanding in terms of connectivity requirements, 3 examples are evoked in this article: lane merge, see through and vulnerable road user protection.

2 Use Case Description and Connectivity Needs

a. **Lane merge**

A vehicle has the intention to merge from one road to another. The vehicle preparing the merge and the one in the main lane can share their local awareness and driving intentions. Based on information and data fusion techniques, trajectories can be computed and then proposed to the vehicles involved. Thereafter, the driving trajectories and speeds of both vehicles can be adjusted, so that the subject vehicle merges into the intended lane in a safe, efficient, and smooth way. Note that legacy vehicles can also be involved, even though they are not connected and thus do not have wireless communication capabilities.

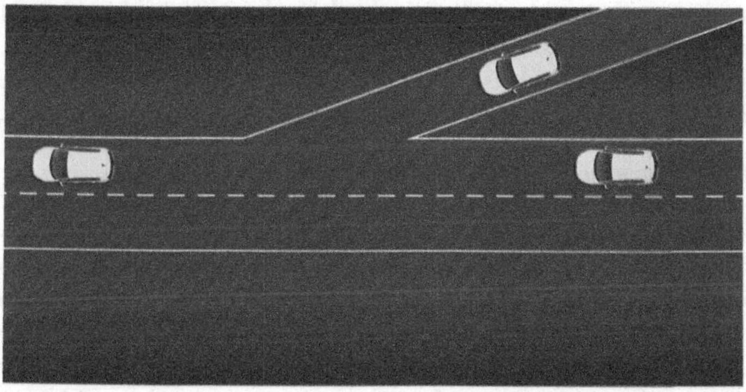

For this use case, two main elements are required from the connectivity perspective: a low latency for all the exchanges between different vehicles, the infrastructure, the cameras helping in the object detection, etc. and the computation capabilities near the road in order to fusion all this information and provide the right instruction to the involved vehicles in the maneuver.

b. **See through**

A cooperative perception system exploits the exchange of data objects detection or video between vehicles via wireless communication to increase safety by facilitating automated driving during maneuvers. The data representing the scene in front of the vehicle ahead of a line is captured by a camera vision system and transferred to the rear vehicle to allow it to see through the forward vehicle and bypass the occluded area. As represented in the figure it will allow for example safe overtaking maneuver.

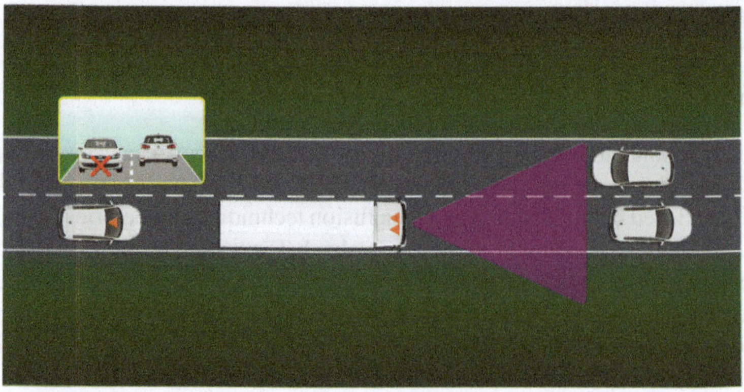

The key connectivity element here is the throughput. The video stream has to be taken into account depending on the image quality and it will challenge the network infrastructure based on the density of video sources coming from different vehicles and/or off-board cameras installed in the road environment such as intersections.

c. **Vulnerable road user protection**

This use case is focused on situations where a pedestrian is moving close to the road or crossing the street. A vehicle is coming down the street with its view hidden behind parked cars. Thanks to the exchange of local information between users via wireless communications, the overall system will determine VRU positioning based on cellular radio signals, GNSS or sensor/camera data. All this location information will be processed from multiple users for alert generation to AD vehicles with high accurate positioning.

The main requirement in this situation is the location accuracy in a urban environment where GNSS may not be precise enough and where the pedestrian has his smartphone anchored to the mobile network. An information aggregator defined as a location server must be placed near the network infrastructure in order to process the radio signal information obtained from different antennas to get a precise positioning information combine it with other sources and then distribute it to the vehicles in the environment.

3 Architecture and Key Technology Enablers

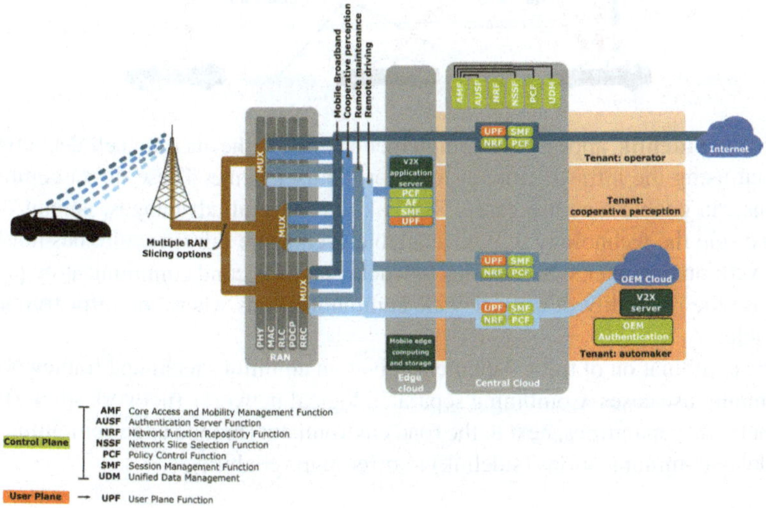

With the arrival of 5G, the core architecture has been completely redefined. The control plane is centered around services rather than interfaces. Instead of entities like it was defined in LTE, now the definition is based on Network Functions. One of the key enablers arriving to the core is the network slicing. **Network Slicing**, is the possibility to have "various simultaneous and different instances of a communication network" running on top of a common infrastructure. This feature shall help to boost new services or enhance the existing ones, by reducing the infrastructure costs and improving the flexibility to cope with dynamic demands from services. This technology defines a structured environment where high bandwidth use cases such as enhanced mobile broadband can be fulfilled without compromising automated driving use cases.

Mobile Edge Computing is another important technology of 5G, consists of installing and operating computing capabilities closer to the final user of a radio communication technology. Having computation capabilities closer to the user reduces latency, improves reliability, and off-loads the core and transport networks. This solves some of the needs evoked in the previous use case definition where an off-board intelligence is needed for data fusion to compute localization and maneuver information from different sources.

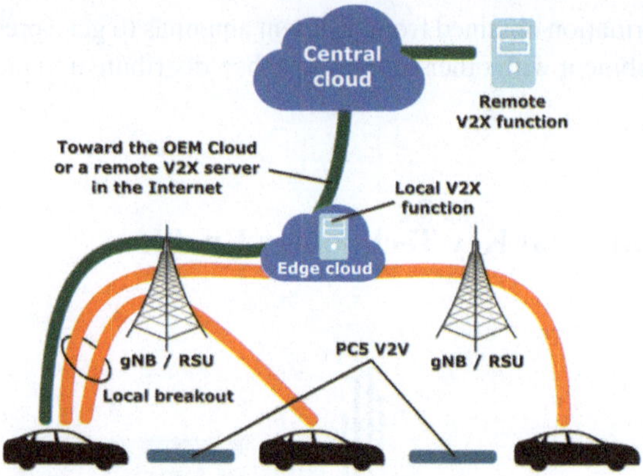

Finally **Sidelink**, appears as a complementary link to the operator cellular network, without using the infrastructure at least for the user plane. These direct communications can enable new use cases. Sidelink, has several advantages, among other: on one side the technology will be available even if the infrastructure has not been deployed; on the other side provides low delay and a second communication path to increase the reliability of critical V2X communications where the infrastructure is available.

The combination of these 3 elements gives an ambitious technical framework for demanding use cases. Combining separated logical networks (network slicing) with computation capabilities, next to the road environment (mobile edge computing) and low delay communications (sidelink) to offer high reliability.

4 Summary

The possibilities of 5G are very promising, but there are still important challenges to solve in order to succeed with the technology deployment for the automotive usages:

- The new technology enablers must be interoperable between different telecom vendors, MNOs and OEMs.
- To provide a seamless experience for automotive services, global roaming is a must, much more in the case of the European Union.
- Frequency harmonization is needed not only in Europe but also worldwide as much as possible to avoid equipment divergences which will increase the user equipment prices.
- V2X as a social benefit, everybody needs to contribute to the deployment of these new services while high prices for frequency band allocation may become a penalty for V2X deployment. The V2X will improve road safety, traffic efficiency and contribute to automated driving as an additional sensor.

- From a regulatory point of view, spectrum allocation should be technology neutral. This applies both to unlicensed spectrum, as for the EU 5.9 GHz ITS band and to licensed spectrum.
- To promote 5G deployment, the infrastructure roll out is required from different actors: public authorities, private investors,…, as it is planned in China market from 2020 on.

Literature

Consideration on spectrum for 5G (2016) CHANG Ruoting Bureau of Radio Regulation, MIIT, China November 9, 2016. https://5g-ppp.eu/wp-content/uploads/2016/11/03_9-Nov_Session-2_Chang-Ruoting-1.pdf
ETSI (2016) "C-ITS Platform: final report"
G white book from China
GCAR: D2.1 Scenarios, use cases, requirements and KPis. Available at https://5gcar.eu/?page_id=97
GCAR: D2.2 Intermediate report on V2X business models and spectrum. Available at https://5gcar.eu/?page_id=97
GCAR: D4.1 Initial design of the 5G V2X system level architecture and security framework. Available at https://5gcar.eu/?page_id=97
GPP TR 22.885 (2015) Study on LTE support for Vehicle-to-Everything (V2X) services
GPP TS 22.185 (2016) Service requirements for V2X services
GPP TR 22.886 (2016) Study on enhancement of 3GPP support for 5G V2X services
GPP TS 22.186 (2017) Service requirements for enhanced V2X scenarios
Wuxi connected city: National Intelligent Connected Vehicle (Shanghai) Pilot Zone

Empowering the Future Mobility

Aleksandar Momcilovic

Abstract Future Mobility enabled by Autonomous driven vehicles, ITS (intelligent transportation system), 5G and V2X will bring more comfort, convenience while travelling, but in first place Safety. Most significant is accident avoidance with 96%. A lot of saved lives and material damages will be the benefit for the whole society. Huawei strives to contribute in many segments of future mobility. As a traditionally communication vendor Huawei is market leader in 5G and V2X, but with recent release of AI computing portfolio including processing brain for autonomous driving called MDC (mobile data center) has opened a new chapter by contributing directly in Automotive industry.

Keywords 5G · C-V2X · ADAS—advanced driver assistance system · ITS (intelligent transportation system)

1 Introduction

We are all experiencing era of new mobility concepts like CarSharing (DriveNow, Car2Go, Free2Move etc.) as well as Car-Hailing like UBER etc. This new approach of especially urban mobility are bringing new economy, new business models for service providers, OEMs and many new players in the emerging Ecosystem.

These all has to be supported by technology, which is enabler of smart mobility concepts.

Connected Car Platform, ITS, V2X, ADAS are crucial technologies enabling new mobility services

A. Momcilovic (✉)
West European OpenLab (Munich OpenLab), Huawei Enterprise Business Group, Huawei Technologies Duesseldorf GmbH, Riesstr. 14, 80992 Munich, Germany
e-mail: Aleksandar.Momcilovic@huawei.com

J. Langheim (ed.), *Electronics Components and Systems for Automotive Applications*, Lecture Notes in Mobility,
https://doi.org/10.1007/978-3-030-14156-1_20

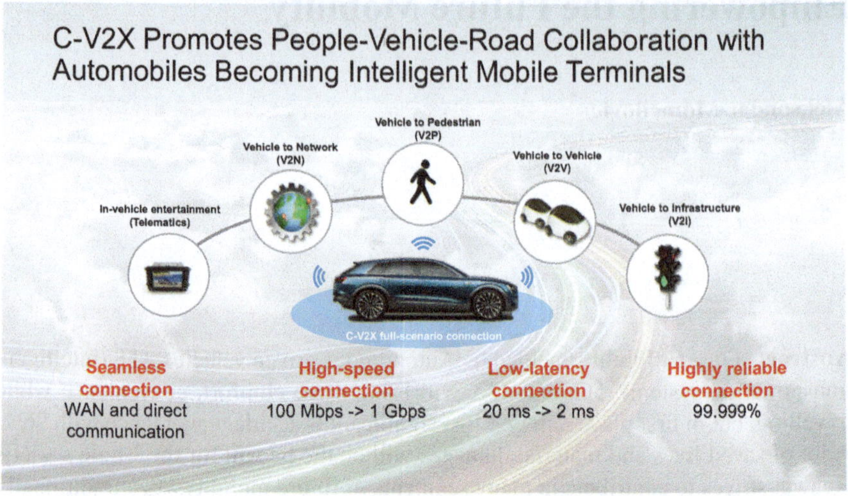

Besides enabling new mobility services to the citizens, collaborative work of those technologies provide additional benefits to the community.

In the first place is Accident Avoidance, where V2X&ADAS collaboratively reduce 96% of traffic accidents.

Followed by reduced Traffic Congestion, Pollution, Transportation Costs and Energy Savings, community will have multi-folded benefits.

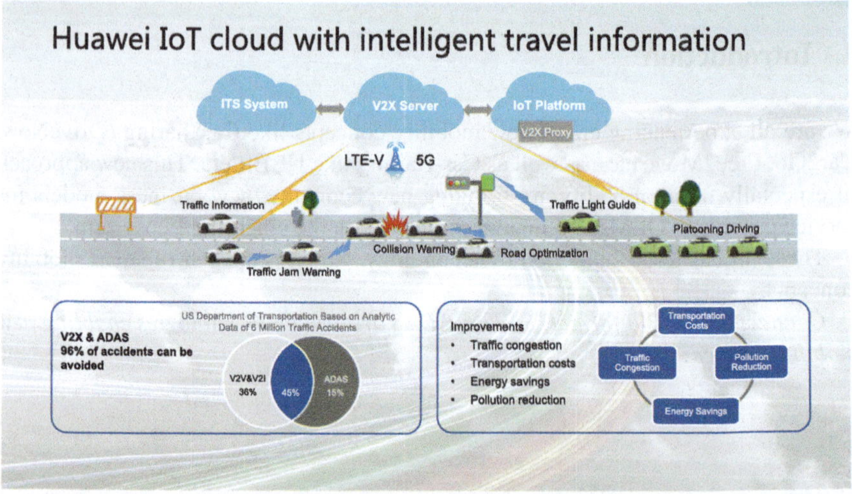

2 Seamless Collaboration of People-Vehicle-Road: A Future Trend

C-V2X promotes People-Vehicle-Road Collaboration with Automobiles Becoming Intelligent Mobile Terminals

In order to get advantages of V2X technology there must be full compliance among device in-car, on the road, handheld terminals.

This turns are back to the topic of selecting V2X radio technology Cellular V2X (4G, 5G) or DSRC short-range WiFi. This raises a clear consideration that V2X has to be regulated and standardized, especially due to the fact that it is primarily about safety. Either one or both as a must can be final decision.

Collaboration of C-V2X and ADAS

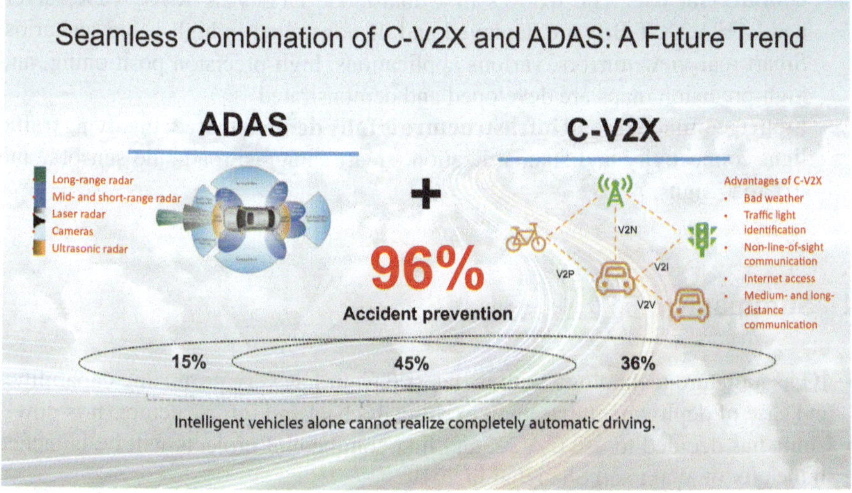

Advantages of C-V2X:

- Bad weather
- Traffic light identification
- Non-line-of-sight communication
- Internet access
- Medium- and long-distance communication

3 C-V2X: Today's Reality!

Wuxi pre-commercial project 2017–2018

Wuxi project achievements:

Current phase is recognized by the industry as the best around the world

- **Largest scale**: 240 intersections, 170 km^2, 5 fast roads, and 1 highway
- **Open data and largest number of scenarios**: Wuxi's traffic police department opens more than 40 types of traffic control information and uses the LTE-V2X technology to provide 12 categories of V2I, V2P, and V2 V information services in 26 application scenarios.
- **Largest, most comprehensive industry ecosystems**: a C-V2X industry ecosystem involving 6 core participants, 23 partners, 10 vehicle enterprises within and outside China, 3 map navigation application developers, 3 smart rear view mirror makers, various applications, various peripheral industry partners, and users from various fields, such as buses, first aid, firefighting trucks, conference reception vehicles, consumer users, and test vehicles of vehicle enterprises
- **Most advanced technologies**:
  - **Communication products are fully verified and are ready for large-scale commercial use**: The world's first dual-mode LTE-V2X RSU, V2X server, and LTE-V2X T-Box are deployed and demonstrated in high-value scenarios. Smart rear view mirrors, various applications, high-precision positioning, and high-precision maps are developed and demonstrated.
  - **Smart, connected road infrastructure is fully demonstrated**: involving traffic light connectivity and standardization, smart cameras, road-side sensors, and IoT road signs.

4 Summary

- 4G as a mature technology with its latest flavour has very promising capabilities and ease of deployment, based on existing 4G backend infrastructure. It is now!
- China has decided to use C-V2X and first commercial projects will be launched in the city of Wuxi and Shanghai in 2019
- Worldwide we already had a lot of trials and POCs of C-V2X and 2019 will be year of real breakthrough
- Smart mobility is part of the Smart City concepts and local communities are interested in investing and as a result there are several trials in European cities e.g. Berlin. They verify both C-V2X and short range DSRC technology.
- Europe Commission has decided to support C-V2X (based of 5G) as a technology for main road infrastructure like Highways and main federal roads. Budget and money is already approved—10 bill EUR will be used as of 2021 to deploy 5G C-V2X along 10.000 km of European Highways.

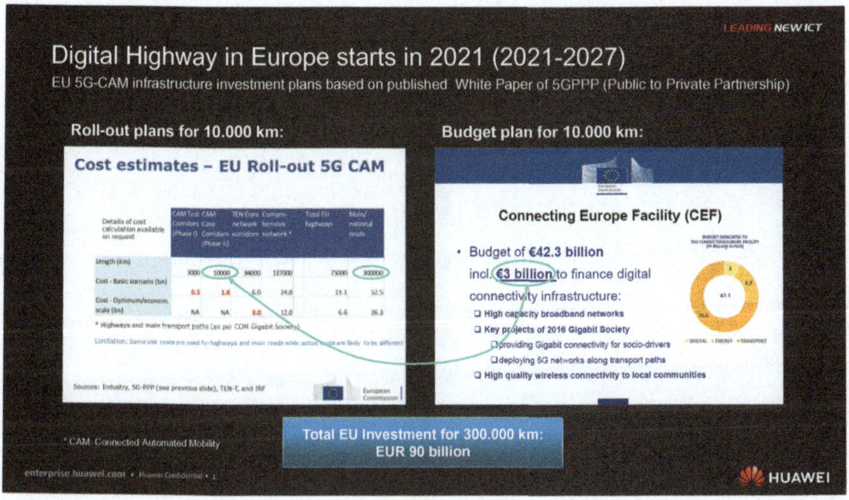

Literature

Huawei project sources
GAA published documents

The GDPR and Its Application in Connected Vehicles—Compliance and Good Practices

Félicien Vallet

Abstract Symbol of the 20th century economy, the automobile is one of the mass consumer products that has impacted society as a whole. Commonly associated with the notion of freedom, cars are often considered as more than just a mean of transportation. Indeed, they represent a private area in which people can enjoy a form of autonomy of decision, without encountering any external interferences. Today, as connected vehicles move into the mainstream, such a vision no longer corresponds to the reality. In-vehicle connectivity is rapidly expanding from luxury models and premium brands to high-volume midmarket models, and vehicles are now massive data hubs. Not only vehicles, but drivers and passengers are also becoming more and more connected. As a matter of fact, many models launched over the past few years on the market integrate sensors and connected on board equipment, which may collect and record, among other things, the engine performance, the driving habits, the locations visited, and potentially even the driver's eye movements, or other biometric data for authentication purposes. Such data processing is taking place in a complex ecosystem, which is not limited to the traditional players of the automotive industry, but is also shaped by the emergence of new players belonging to the digital economy. Aware of the issues at stake for the protection of motorists' privacy in this ecosystem, the Commission Nationale de l'Informatique et des Libertés (CNIL)—the French data protection authority—developed a reference framework enabling professionals to comply with the General Data Protection Regulation (GDPR), applicable as from 25 May 2018, thus making the compliance simpler and ensuring that users enjoy transparency and control in relation to their data.

Keywords Data protection · Privacy · Connected vehicle · Personal data · General data protection regulation (GDPR)

The views and opinions expressed in this article are those of the author and do not necessarily reflect the official policy or position of the CNIL.

F. Vallet (✉)
Commission Nationale de l'Informatique et des Libertés (CNIL), 3 place de Fontenoy, TSA 80715, 75334 Paris Cedex 07, France
e-mail: fvallet@cnil.fr

J. Langheim (ed.), *Electronics Components and Systems for Automotive Applications*, Lecture Notes in Mobility,
https://doi.org/10.1007/978-3-030-14156-1_21

1 Introduction

According to the Comité des Constructeurs Français d'Automobiles (CCFA) and the European Automobile Manufacturers' Association (ACEA), the French and European car fleets are respectively of 39 and 299 million entities in 2017 (European Automobile Manufacturers' Association (ACEA) 2018, Comité des Constructeurs Français d'Automobiles 2018). At this scale the development of the next generation of vehicles is a major industrial issue and a Statista study predicts that the number of vehicles with connectivity capacities is expected to amount to 106 Million in Europe in 2023 (respectively 15 Million in France) (Statista Connected cars in Europe 2018).

It is commonly recognized that automated and connected vehicles can offer significant benefits for users as well as for the general public. Whether for road safety, maintenance, eco-driving or for personalised motor insurance, the capacity for innovation around the vehicle is considerable as illustrated in Fig. 1. The rapid advancement of connected vehicle technologies allowing the development and introduction of such new and innovative products, devices or telematics services, necessitates the collection and processing of massive amounts of data. These data are collected by a wide variety of sensors installed inside and outside of the vehicle and are susceptible to be communicated to a great number of actors: vehicle and equipment manufacturers, insurers, car dealers, garages but also with vehicles circulating nearby or road infrastructure equipment (Future of Privacy Forum (FPF) 2017). On top of this, it turns out that a large chunk of these data are personal data, which raises the stakes of their processing.

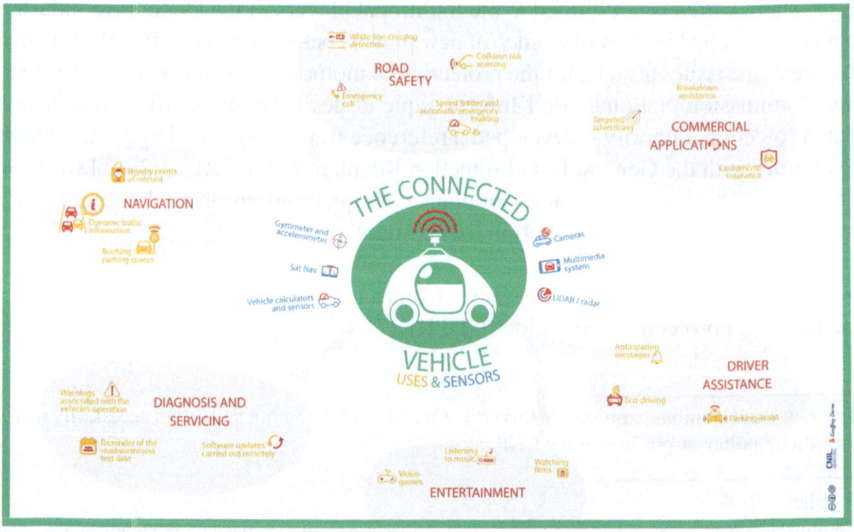

Fig. 1 The connected vehicle—uses and sensors

In 2016, the Fédération Internationale de l'Automobile (FIA) ran a campaign across Europe called "My Car My Data" to get a sentiment on what Europeans think about connected cars (Fédération Internationale de l'Automobile (FIA) 2015). While it showed the high interest of drivers for connectivity, it also highlighted the vigilance that must be exercised with regard to the use of the data produced by vehicles as well as the importance of complying with personal data protection legislation. Thus, the challenge is, for each stakeholder, to incorporate the protection of personal data "dimension" from the product design phase, and to ensure that users enjoy transparency and control in relation to their data. Such an approach helps to strengthen user confidence, and thus the long-term development of those technologies.

2 Privacy and Data Protection Risks

As previously outlined, it is widely acknowledged that the stakes are high regarding data protection, privacy and cybersecurity of the connected vehicles. Indeed drivers and their passengers are—through their vehicles—increasingly connected and numerous risks can be identified.

Lack of control and transparency

Vehicle drivers and passengers may not always be adequately informed about the processing of data taking place in or through a connected vehicle. The information may be given only to the vehicle owner, who may not be the driver, and may also not be provided in a timely fashion. This point is of importance since, during their lifetime, vehicles may belong to more than one owner either because they are sold or because they are being leased rather than purchased. In addition, vehicles are increasingly being shared or rented, not just by companies, but also individuals, and the person whose data is collected may not be able to object to some data processing as detailed by the NGO Privacy International (Privacy International 2017).

Unlawful processing

Data collected through connected vehicles for well-defined purposes might later be used or sold for unconsented or unauthorised ones. Furthermore, data collected by connected vehicles could be used by law enforcement authorities to detect speeding or other infractions, or for surveillance. In this case such data will be considered as relating to criminal convictions and offences.

Excessive collection

With the ever increasing number of sensors being deployed in connected vehicles there is a very high risk of excessive data collection compared to what is necessary. The development of new functionalities and more specifically those based on machine learning algorithms may require a large amount of data collected over a long period of time.

Limitation on the possibility to remain anonymous

Connected vehicles also raise significant concerns regarding the processing of location data as its intrusive nature can put a strain on the current possibilities to travel anonymously. Location data is indeed particularly revealing since the journeys carried out are very characteristic in that they enable to infer the place of work and of residence, as well as the driver's centres of interest (e.g., leisure, and, possibly, religion through the place of worship, or sexual orientation through the places visited).

Security of personal data

The plurality of functionalities, services and interfaces (e.g., web, USB, RFID, Wi-Fi) offered by connected vehicles increases the attack surface and thus the number of potential vulnerabilities through which personal data could be compromised. Unlike most Internet of Things devices, connected vehicles are critical systems where a security breach may endanger the life of its users and people around. In addition, personal data stored on vehicles and/or at external locations (e.g., in cloud computing applications) may not be adequately secured against unauthorized access.

3 Related Works

Connected vehicles have become a substantial subject for data protection regulators over the last decade, with a major increase in the last couple of years. Various works have thus been published at the national and international levels concerning the security and privacy risks connected vehicles expose their drivers and passengers to. Those regulations and initiatives aim at complementing the existing data protection and privacy frameworks and providing guidance to professionals.

On the European level, various initiatives related to the connected vehicles have been discussed by regulators, in some cases for quite some times. For instance, from 31 March 2018 onward, a 112-based eCall in-vehicle system is mandatory on all new types of M1 and N1 vehicles (passenger cars and light duty vehicles) in Europe (The interoperable EU-wide eCall 2015; Decision N° 585/2014/EU of the European Parliament 2014). In 2006, the Article 29 Working Party had already adopted a working document on data protection and privacy implications in eCall initiative (Article 29 Working Party 2006). More recently, in 2017, the Article 29 Working Party adopted an opinion regarding the processing of personal data in the context of Cooperative Intelligent Transport Systems (C-ITS) (Article 29 Working Party 2014). In addition to this, the European Union Agency for Network and Information Security (ENISA) published a study focused on cybersecurity and resilience of smart cars listing the sensitive assets as well as the corresponding threats, risks, mitigation factors and possible security measures to implement (European Union Agency for Network and Information Security (ENISA) 2017). On the international level, various gathering of data protection authorities recently adopted resolutions and working papers on connected vehicles such as the International Conference of Data Protection and Privacy Commissioners (ICDPPC) (International Conference of Data Protection and Privacy Commissioners (ICDPPC) 2017) and the International Working Group on Data Protection in Telecommunications (IWGDPT) (International Working Group on Data Protection in Telecommunications (IWGDPT) 2018).

On the national level, various initiatives were also recorded recently, such as the common declaration on the principles of data protection in connected and not-connected vehicles published in 2016 by the Conference of the German Federal and State Data Protection Authorities and the German Association of the Automotive Industry (VDA) (Conference of the German Federal and State Data Protection Authorities and the German Association of the Automotive Industry (VDA) 2016). In order to raise awareness on the cybersecurity matter and to provide voluntary guidance to encourage best practices and prioritizes safety, various guides and papers were released such as those provided by the UK Centre for Connected and Autonomous Vehicles (CCAV) (UK Centre for Connected and Autonomous Vehicles (CCAV) 2017) and the U.S. National Highway Traffic Safety Administration (NHTSA) (National Highway Traffic Safety Administration (NHTSA) 2017; National Highway Traffic Safety Administration (NHTSA) 2018). Finally, in October 2017, the French data protection authority, the Commission Nationale de l'Informatique et des Libertés (CNIL), released a compliance package for connected cars in order to provide assistance to stakeholders on how to integrate data protection by design and by default, enabling data subjects to have effective control over their data (Commission Nationale de l'Informatique et des Libertés (CNIL) 2017).

4 Principles of the GDPR

Entered into force on May 25, 2018, the General Data Protection Regulation (GDPR) (EU General Data Protection Regulation (GDPR) 2016) provides a new set of rules for the processing of personal data. The objective is to give citizens back control over of their personal data, and to simplify the regulatory environment for business through a unified framework over the European continent. This data protection reform is viewed as a key enabler of the digital single market which has been designed to allow European citizens and businesses to fully benefit from the digital economy. In particular, the GDPR lays out responsibilities for organisations to ensure the privacy and protection of personal data, provides data subjects with certain rights, and assigns powers to regulators to ask for demonstrations of accountability or even impose fines in cases where an organisation is not complying with the GDPR requirements. In order to understand these requirements, a few key principles can be highlighted:

- *Lawfulness, fairness and transparency*: Organisations that process personal data such as companies, public authorities, associations, etc. are asked to do so in a in a lawful, fair and transparent manner for the concerned data subjects.

- *Limitation of purpose, data and storage*: Organisations are expected to limit the processing, collect only that data which is necessary, and not keep personal data once the processing purpose is completed.
- *Data subject rights*: The data subjects have been assigned the right to ask the data processor what information it has about them, and what it does with this information. In addition, a data subject has the right to ask for correction, object to processing, lodge a complaint, or even ask for the deletion or transfer of his or her personal data.
- *Privacy by design and by default*: Organisations must incorporate organisational and technical mechanisms to protect personal data from the design stage of their systems and processes. In addition, once a product or service has been released to the public, the strictest privacy settings should apply by default, without any manual input from the end user.
- *Data integrity and confidentiality*: Organisations processing personal data must put in place appropriate technical or organisational measures in a manner that ensures security of the personal data, including protection against unauthorised or unlawful processing and against accidental loss, destruction or damage, using appropriate technical or organisational measures.
- *Accountability*: Under the GDPR, organisations are responsible for their actions and must be able to demonstrate their compliance with the data protection obligations. For this they can rely on the Data Protection Impact Assessment (DPIA), a new tool whose goal is to systematically analyse, identify and minimise the data protection risks of a new project or product.

This new regulation should therefore not only be considered as a technical or legal project. It also offers many opportunities to companies that can take advantage of it to secure and protect all their data, as well as to accelerate their digital transformation. However, in order to be efficiently applied to a given economical sector, the GDPR principles must be derived according to its specificities.

5 Application to the Automotive Ecosystem

The GDPR applies to the processing of personal data. In the case of the automotive industry, personal data include all data associated or that can be associated with a natural person (driver, vehicle owner, etc.), notably via the vehicle serial number or the vehicle licence plate number. They may be directly identifying data, such as the driver's name, as well as indirectly identifying data (e.g. details of journeys made, data relating to the driving style or the distance covered), or the vehicle's technical data (e.g. data relating to the wear and tear on vehicle parts), which, by cross-referencing with other files, can be related to a natural person. It covers therefore a very broad spectrum of data.

As stated above, in March 2016, the Commission Nationale de l'Informatique et des Libertés (CNIL) launched a consultation process with automotive professionals (vehicle and equipment manufacturers, innovative service providers, rental companies, etc.), as well as insurance and telecoms professionals and the French public sector (Gendarmerie Nationale, transport ministry, agency for the environment and energy management, etc.). This approach has led to the publication of a "compliance package" (Commission Nationale de l'Informatique et des Libertés (CNIL) 2017), providing best practices in the collection and use of personal data by connected vehicle to the professionals.

General recommendations

In order to mitigate the risks for drivers and passengers, deployed systems must be configured to respect the privacy of individuals by design and by default as stated in the GDPR. For this, various recommendations should be followed by professionals involved in the automotive ecosystem.

First, technologies should be designed to minimize the collection of personal data, provide privacy-protective default settings and ensure that data subjects have the option to easily modify configurations associated with their personal data. For instance, vehicle and equipment manufacturers, service providers and other data controllers should, wherever possible, use processes that do not involve transferring personal data outside of the vehicle (i.e., the data is processed internally). Such a local processing enables to deal with sensitive data such as biometric data or data relating to criminal offenses or other infractions, as well as detailed location data which otherwise would be subject to stricter rules.

It is also essential to raise awareness of the principles of informational self-determination, transparency and fair collection, which involve, at a minimum, informing the persons concerned and even obtaining their consent in some cases. In practice, these requirements can be translated into the implementation of technical devices, such as ergonomic and easily configurable dashboards, to guarantee users control and monitoring of their data. Drivers must therefore be able to easily suspend and configure the uses made of their data. Moreover, the GDPR recommends that professionals in the same economic sector adopt standardized icons to provide users with an overview that is clearly understandable and the automotive industry seems perfectly suited for such an initiative.

Finally, vehicle and equipment manufacturers, service providers and other data controllers also must put in place measures that guarantee the security and confidentiality of processed data and take all useful precautions to prevent control being taken by an unauthorised person. In particular, industry participants should consider adopting measures such as:

- The encryption of the communication channels by means of a state-of-the-art algorithms;
- The use of encryption-key management systems unique to each vehicle (and not to each model)
- The encryption of data stored remotely by means of state-of-the-art algorithms;
- The protection of encryptions keys from any disclosure;
- The authentication of data-receiving devices;

– The use of mechanisms ensuring data integrity (e.g., by using hash functions);
– Etc.

Three case scenarios for data circulation

On a data protection and privacy standpoint, the connected vehicles definition should be interpreted in a broad manner. Indeed, personal data can be collected through several means, including: vehicle sensors, telematics boxes or mobile applications (e.g. accessed from a device belonging to a driver). Mobile applications offering driving related services such as GPS navigation, eco-driving advice or safe driving monitoring are enabling the vehicle connectivity capacities even though they may not effectively rely on the transmission of data with the vehicle per se. Therefore, based on the various data flow patterns identified, three case scenarios can be designed as illustrated in Fig. 2:

– *In–In*: the data collected in the vehicle remains under the sole control of the user and is not transmitted outside the vehicle. Examples: eco-driving solution displaying real-time advice to the driver, preventive maintenance alerting the driver to the condition of his vehicle.
– *In–Out*: the data collected in the vehicle are transmitted externally to provide a service to the data subject. Examples: "Pay as you drive" insurance contract, anti-theft device allowing the location of the vehicle, automatic emergency call (eCall).
– *In–Out–In*: the data collected in the vehicle is transmitted to the outside world to trigger an automatic action in the vehicle. Examples: dynamic traffic navigation system, remote modification of the charging capacities of the battery of an electric vehicle, reception of over-the-air technical updates.

As detailed in Commission Nationale de l'Informatique et des Libertés (CNIL) 2017, following these information flow schemes it is then possible to define, for the main categories of use, the intended purposes, the legal basis on which data may be processed, the categories of data collected, the retention periods of such data, the rights

Fig. 2 The three identified data circulation patterns in the connected vehicle: In–In, In–Out and In–Out–In

of individuals, the recipients of this information and the security measures to be implemented.

6 Conclusion and Forthcoming Works

Following the work presented above, European data protection authorities—gathered in the European Data Protection Board (EDPB) by the GDPR—now ambition to publish commonly agreed guidelines in order to empower and secure automotive professionals with good practices regarding data protection and privacy at the level of the European continent. Some topics, not covered by the previous works will also need to be addressed in the future, for instance: the management of private or professional fleets of vehicles, the deployment in Europe of collaborative intelligent transport systems (C-ITS) or the upcoming regulation on privacy and electronic communications (known as the ePrivacy regulation) (Proposal for a Regulation of the European Parliament 2017). The permanent evolution of technologies will also lead the data protection authorities to maintain continuous exchanges with the actors of the connected vehicle ecosystem in order to ensure the supervision of the new uses that will be made of the personal data of the drivers and of their passengers.

References

Article 29 Working Party—Working document on data protection and privacy implications in eCall initiative—2006; https://ec.europa.eu/justice/article-29/documentation/opinion-recommendation/files/2006/wp125_en.pdf

Article 29 Working Party—WP252, Opinion 03/2017 on processing personal data in the context of Cooperative Intelligent Transport Systems (C-ITS)—2017; http://ec.europa.eu/newsroom/just/document.cfm?doc_id=47888

Comité des Constructeurs Français d'Automobiles—L'industrie automobile française—Analyse et statistiques—2018; http://ccfa.fr/wp-content/uploads/2018/09/analyse_statistiques_2018_fr.pdf

Commission Nationale de l'Informatique et des Libertés (CNIL)—Compliance package for a responsible use of data in connected cars—2017; https://www.cnil.fr/sites/default/files/atoms/files/cnil_pack_vehicules_connectes_gb.pdf

Conference of the German Federal and State Data Protection Authorities and the German Association of the Automotive Industry (VDA)—Data protection aspects of using connected and non-connected vehicles—2016; https://www.lda.bayern.de/media/dsk_joint_statement_vda.pdf

Decision N° 585/2014/EU of the European Parliament and of the Council of 15 May 2014 on the deployment of the interoperable EU-wide eCall service Text with EEA relevance; https://eur-lex.europa.eu/legal-content/EN/TXT/PDF/?uri=CELEX:32014D0585

European Union Agency for Network and Information Security (ENISA)—Cyber security and resilience of smart cars—2017; https://www.enisa.europa.eu/publications/cyber-security-and-resilience-of-smart-cars

EU General Data Protection Regulation (GDPR): Regulation (EU) 2016/679 of the European Parliament and of the Council of 27 April 2016 on the protection of natural persons with regard to the processing of personal data and on the free movement of such data, and repealing Directive 95/46/EC (General Data Protection Regulation), OJ 2016 L 119/1; https://eur-lex.europa.eu/legal-content/EN/TXT/?uri=CELEX%3A32016R0679

European Automobile Manufacturers' Association (ACEA)—Vehicles in use—2018; https://www.acea.be/statistics/tag/category/vehicles-in-use

Fédération Internationale de l'Automobile (FIA)—Campaign "My Car My Data"—2015; http://www.mycarmydata.eu/

Future of Privacy Forum (FPF)—Infographic "Data and the connected car" 2017; https://fpf.org/wp-content/uploads/2017/06/2017_0627-FPF-Connected-Car-Infographic-Version-1.0.pdf

International Conference of Data Protection and Privacy Commissioners (ICDPPC)—Resolution on data protection in automated and connected vehicles—2017; https://edps.europa.eu/sites/edp/files/publication/resolution-on-data-protection-in-automated-and-connected-vehicles_en_1.pdf

International Working Group on Data Protection in Telecommunications (IWGDPT)—Working paper on connected vehicles—2018; https://www.datenschutz-berlin.de/fileadmin/user_upload/pdf/publikationen/working-paper/2018/2018-IWGDPT-Working_Paper_Connected_Vehicles.pdf

National Highway Traffic Safety Administration (NHTSA)—Automated Driving Systems 2.0: A Vision for Safety—2017; https://www.nhtsa.gov/sites/nhtsa.dot.gov/files/documents/13069a-ads2.0_090617_v9a_tag.pdf

National Highway Traffic Safety Administration (NHTSA)—Automated Vehicles 3.0: Preparing for the Future of Transportation—2018; https://www.transportation.gov/sites/dot.gov/files/docs/policy-initiatives/automated-vehicles/320711/preparing-future-transportation-automated-vehicle-30.pdf

Privacy International—Connected Cars: What Happened to Our Data on Rental Cars—2017; https://privacyinternational.org/sites/default/files/2017-12/cars_briefing.pdf

Proposal for a Regulation of the European Parliament and of the Council concerning the respect for private life and the protection of personal data in electronic communications and repealing Directive 2002/58/EC (Regulation on Privacy and Electronic Communications), 2017; https://eurlex.europa.eu/legal-content/EN/TXT/?uri=CELEX:52017PC0010

Statista—Connected cars in Europe—2018; https://www.statista.com/outlook/320/102/connected-car/europe

The interoperable EU-wide eCall 2015; https://ec.europa.eu/transport/themes/its/road/action_plan/ecall_en

UK Centre for Connected and Autonomous Vehicles (CCAV)—2017; Principles of cyber security for connected and automated vehicles; https://www.gov.uk/government/publications/principles-of-cyber-security-for-connected-and-automated-vehicles.

The GDPR and Its Application in IoT and Connected Cars Opportunities for Business and Competitivity

Gaëlle Kermorgant and Michèle Guilbot

Abstract The General Data Protection Regulation (GDPR) is in force since May 25 2018. This has an important impact on the design and development of new technologies based on exploitation of data, in particular private data. Not complying with the requirements can have high financial consequences up to 4% of the worldwide turnover of a company as a fine. At first glance, GDPR seems to be a constraint. However the, this regulation offers also the opportunity to develop new services based on big data processing and become competitive in a domain, which is considered being the petroleum of the 21st century. The paper describes the impact and opportunities of the GDPR on the Internet of things (IoT) and connected cars by highlighting two examples:

- the first aims at improving road safety by calculating a risk factor based on analysis of driver behaviour, data collection and artificial intelligence
- the second aims at improving road infrastructure by detecting road deficiencies or risk zone through in car data collection.

Keywords Personal data protection · Connected cars · IoT · GDPR · Privacy

1 Introduction

The mobility sector faces major challenges, in particular the fight against pollution, the reduction of accidents and regulation of road traffic. Information and communication technologies (ICTs) can partly meet these challenges.

G. Kermorgant (✉)
Paris Bar, Paris, France
e-mail: gmkermorgant@gmail.com

M. Guilbot
Accident Mechanisms Laboratory, IFSTTAR, French Institute of Science and technology for Transport, Development and Network, Paris, France
URL: https://www.ifsttar.fr/en/welcome/

J. Langheim (ed.), *Electronics Components and Systems for Automotive Applications*, Lecture Notes in Mobility,
https://doi.org/10.1007/978-3-030-14156-1_22

In addition, they are a growing opportunity for companies that will invest in these areas. The control of data (personal or not) collected by road infrastructure, connected vehicles and everyday objects is one of the tools for the development of intelligent transport services (ITS).

Nevertheless, in a connected road traffic system, in-vehicle devices, either mandatory or optional, must offer the best possible guarantees of integrity and digital security to respect freedom of drivers to travel anonymously, to protect confidentiality of users' personal data, to prevent risk of unlawful intrusion, to ensure the availability and quality of information provided to drivers and automated equipment in vehicles.

In the European Union "*The protection of natural persons in relation to the processing of personal data is a fundamental right. Article 8(1) of the Charter of Fundamental Rights of the European Union (the 'Charter') and Article 16(1) of the Treaty on the Functioning of the European Union ('FEU) provide that everyone has the right to the protection of personal data concerning him or her*". This protection, reaffirmed by the GDPR (recital 1),[1] is also taken into account by various texts, for example a draft Regulation under discussion on the protection of privacy in communications systems and value-added services.[2] National legislation and practices must take these developments into account.[3]

In parallel with the activities of the legislator and noting that connected vehicles are developing at high speed, the French Data Protection Authority, the National Commission for Computing and Liberties (CNIL) published in October 2017, a "Connected vehicles and personal data" compliance package (Compliance Package) to help companies to propose solutions that obey right to the protection of personal data.[4] The latter having is presented by Félicien Vallet from the CNIL, we will not go back on its content.[5] Nevertheless, it should be recalled that the Compliance Package specifies that personal data include geolocation data,[6] technical data related to the condition of the vehicle and parts, biometric data of the driver, data related to the use of the vehicle by the driver or passengers.

---

[1]GDPR 2016/679 which follow the Directive 95/46/EC of the European Parliament and of the Council of 24 October 1995 on the protection of individuals with regard to the processing of personal data and on the free movement of such data.

[2]REGULATION OF THE EUROPEAN PARLIAMENT AND OF THE COUNCIL concerning the respect for private life and the protection of personal data in electronic communications and repealing Directive 2002/58/EC (Regulation on Privacy and Electronic Communications) which will replace the DIRECTIVE 2002/58/EC OF THE EUROPEAN PARLIAMENT AND OF THE COUNCIL of 12 July 2002 concerning the processing of personal data and the protection of privacy in the electronic communications sector (Directive on privacy and electronic communications).

[3]See in particular for France the revision of the Act n°78-17 of the 6th January 1978 amended by Act n°2018-493 of the 20th June 2018.

[4]Compliance Package «Connected vehicles and personal data». https://www.cnil.fr/sites/default/files/atoms/files/cnil_pack_vehicules_connectes_gb.pdf.

[5]Presentation of Félicien VALLET: GDPR—general introduction.

[6]This clarification had already been stated in Resolution (délibération) 2006-066 of the 16th March 2016 adopting a recommendation on the implementation of devices to geolocate motor vehicles used by employees of a private or public body. See also GDPR, art.4.1.

Therefore, all regulations relating to protection of personal data apply to technologies that make it possible to identify a natural person directly or indirectly.

We extend here the theoretical presentation proposed by the CNIL with two illustrations showing the possible ways of implementing innovations in the connected vehicle that obey the protection of drivers' personal data. These illustrations are intended to demonstrate that regulation does generate constraints, but it does not hinder the development of innovative products. It also allows them to offer the best possible security guarantees and its compliance could ultimately constitute a competitive advantage for economic actors by promoting user's confidence.

The first illustration aims at preventing road risk in driving situations (Illustration of an IN-IN application according to the CNIL Compliance Package). Under this scenario, personal data remains inside the vehicle, is not transmitted to a third party and remains under user's control (e.g. eco-driving).

The second illustration also aims at preventing road risk by offering partner road managers an auscultation of their network by processing data collected thanks to event data recorder in service vehicles in their fleet (Illustration of an IN-OUT application according to the CNIL Compliance Package). Under this scenario, personal data is transmitted to a third party for the provision of service(s) to the user. For example, for commercial exploitation consented by the user on a contractual basis (such as Pay-as-you-drive) or the fight against theft.[7] The programme proposed here provides a service of general interest to users which is not immediate (improvement of road infrastructure) but essential for safety of their trips.

2 A Tool to Prevent Road Risk: The API SafetyNex

Specializing in artificial intelligence and perception of intelligent and connected vehicles, Nexyad has created, among others, the SafetyNex algorithm for real-time driving risk assessment that alerts the driver in case of danger. At the end of a trip made by the user, this application programming interface (API) delivers a score based on the contextualized conduct behaviour of its user. The goal of the treatment is to prevent crashes by alerting the driver to help him adapt his/her driving to his/her environment.

Aware of the issues of protection of personal data/privacy involved in the calculations carried out by their API and in accordance with the principle of "privacy by design", the two creators of the company Nexyad have, from the outset/design of their

[7]A third scenario, IN-OUT-IN, is not illustrated in our article. According to this scenario, data is transmitted outside the vehicle before returning to it, in order to trigger an action. For example, a dynamic navigation system that allows you to send back live information on the state of road congestion in order to calculate a new route.

project, sought to comply with the Data protection Act and the recommendations and deliberations of the CNIL.[8]

This goal first led them to opt for stronger core logical and physical security measures.[9] Indeed, encryption and cryptology are at the heart of the security guarantees of APIs developed by Nexyad. Although this aspect of the API meets the requirements for protection of personal data, we will not elaborate this point. We will focus on the two main challenges faced by SafetyNex regarding the protection of personal data: the processing of the location and speed of the vehicle (2.1) and obtaining informed consent from users of the API as to their geolocation (2.2).

2.1 A Processing that Respects the Freedom to Come and Go Anonymously

Concerning the SafetyNex API, its effectiveness is based on a certain and rigorous geolocation of the driver. Indeed, the ability to accurately locate the vehicle makes it possible to know characteristics of road used (crossing, roundabout, pedestrian crossing, etc.). However, as we have already pointed out, the purpose of processing is not to geolocate users. This data is used to assess driving behaviour in one's environment. Nevertheless, in many occasions, CNIL has stressed the particularly intrusive nature of any device allowing the location of people: permanent monitoring of people must be avoided as far as possible. The infringement of the freedom to come and go anonymously must be clearly justified and proportionate to the pursued objective. It was therefore important for Nexyad to find a solution that would allow users to maintain user freedom.

In addition to geolocation, it is obvious that the API must know the speed of the vehicle user in relation to its location. It is therefore likely to identify possible traffic violations (and in particular non-compliance with speed limits).

However, Section 10 of the GDPR states that "*Processing of personal data relating to criminal convictions and offences (…) shall be carried out only under the control of official authority or when the processing is authorised by Union or Member State law providing for appropriate safeguards for the rights and freedoms of data subjects. Any comprehensive register of criminal convictions shall be kept only under the control of official authority*".

Article 9 of the French Data Protection Act supplements the GDPR on this issue by listing in a restrictive manner categories of people authorised to carry out processing operations intended to directly reveal data relating to offences. This list does not include private legal entities such as service providers such as Nexyad or insurance companies.

[8]See in particular the Decision (délibération) n°2005-278 of the 17th November 2005 refusing the implementation by MAAF Assurances SA of automated processing of personal data based on the geolocation of vehicles.

[9]Op. cit.: French Act n°78-17, article 34.

In order to avoid an unjustified infringement of the freedom to come and go anonymously and to avoid collection of traffic violations, Nexyad's engineers have decided that:

- The geolocation and speed data would be processed directly in the smartphone or computer of the API user.
- Once the calculation is completed, data would be immediately deleted.

Thus, even if the device can be offered by insurance companies as part of a "pay-how-you-drive" contract, insurers will have no access to the insured's personal data as recommended by the CNIL.[10] Only the score obtained by the insured using the API at the end of a trip will be known. This product is therefore in compliance with article 10 of the GDPR versus article 9 of the French Data Protection Act.

It should be recalled that as early as 2010, the CNIL advised that such data "*should be aggregated directly into the box if possible so that no detailed information is sent back to either the service provider or the insurer. Failing this, the CNIL recommends that the data be aggregated by the service provider, with the latter being responsible for sending to the insurer the data already aggregated*".[11] Similarly, the Compliance Package recommends that, as far as possible, vehicle usage data should be processed directly in the telematics boxes, so that the service provider only accesses to the result data (e.g. a score) and not to the detailed raw data.

In addition to the implementation of these measures, which also contribute to data security and confidentiality, Nexyad's engineers had to consider the best ways to obtain informed consent from their API users regarding their geolocation.

2.2 *An Informed Consent of API Users Regarding Their Geolocation*

Reading general terms and conditions of use of digital products can be very tedious and time consuming. In addition to technical language, the number of pages to be read usually leads users to approve them without paying the attention that the protection of their personal data deserves.

In this context, in order to meet the requirements requiring collection of free, informed, specific and unambiguous consent, as recalled by the CNIL in the Compliance Package, Nexyad is developing several ways to bring the processing of vehicle location to the attention of API users:

- First of all, the purchase of this device would be subject to collection of specific consent of the user (at the integrator's expense).

[10] Op. cit.: Decision n°2005-278.

[11] Free translation of the Resolution (Délibération) n° 2010-096 of the 8th April 2010 on the implementation by insurance companies and car manufacturers of in-vehicle geolocation devices.

- In addition, the user should launch the functionality to benefit from the service: the geolocation is not automatically performed by default and/or continuously when the vehicle starts.
- Besides, if the user does not want to be located, he will have the possibility to disable the API at any time.
- Finally, the user is informed of the activation of geolocation by moving an arrow on the smartphone screen that reproduces the environment in which the vehicle is operating, which is certainly one of the best ways for the user to become aware of the processing of his location data.

Therefore, we note that despite the legal constraints to offer a powerful tool for road prevention, the SafetyNex API makes it possible to innovate in this field while complying with the requirements of the GDPR.

3 A Program to Improve the Management of Roads: S_VRAI and DYMOA

In collaboration with Cerema,[12] IFSTTAR is conducting two national projects, S_VRAI and DYMOA,[13] aiming at identifying improvement for road accident prevention based on the use of data collected by on-board units in vehicles.[14] In addition to accident studies, these two projects contribute to improving road infrastructure, particularly for the S_VRAI project, through a partnership with road managers who make public vehicles available to projects by some of their agents during their professional activity.[15]

We have extracted for the presentation the modalities developed by IFSTTAR's Laboratory Mechanisms of Accidents in these projects to respect the main protection rules.[16] Three points will be addressed: the consent of drivers (3.1), the measures taken to avoid their identification (3.2), the data minimization principle illustrated by geolocation (3.3).

3.1 Voluntary Collection and Processing of Data

As CNIL points out, the consent of employees is not always free in the event of hierarchical subordination with the controller. Although IFSTTAR is the controller,

[12] A French Institut Le Centre d'Etudes et d'expertises sur les Risques, l'Environnement, la Mobilité et l'Aménagement est un établissement public administratif français. Adresse site.

[13] S_VRAI: Saving lives through incident analysis feedback. DYMOA: infrastructure diagnosis and vehicle dynamics for motorcycles and cars. For a presentation see the attached appendices.

[14] RDR or Road data recorders.

[15] The collection for DYMOA being carried out using equipment embedded on private vehicles.

[16] On the main measures adopted see Guilbot et al. (2016a, 2016b, 2017, 2018).

and not the road managers, data collection is based on driver voluntarism. In addition, IFSTTAR has undertaken not to allow the managers of the agents to identify them thanks to the restitutions of results which will be aggregated.

Indeed, on the one hand participation in S_VRAI is not an obligation directly related to the professional activity of the agents (in the sense of their job description) and, on the other hand, it must not impact the behaviour of their hierarchy towards them. Control of their participation therefore implies several complementary consequences in terms of technical and organizational measures, in particular:

- collection by IFSTTAR of a written consent of the volunteers;
- default deactivation of vehicles likely to be driven by several drivers to protect non-voluntary drivers from the risk of recording their data;
- reversibility of consent, for voluntary drivers without having to justify it, including in fleet vehicles (public vehicles) withdrawal of the participation may concern a course (temporary deactivation) or be definitive;
- guarantee that the data concerning them will not be transmitted to their hierarchy.

All potential drivers of an equipped vehicle have been informed by IFSTTAR and/or by Cerema about the project, the methods of processing and the recipients of the data (briefing notes, oral presentation of the project). The opinion of the staff representatives for the fleets was collected at a Joint Technical Committee.[17] Volunteers have also been informed that the recorded data may be requisitioned by the judicial authority as an authorized third party, particularly in the event of an accident. This is to obtain very clearly an informed consent.

These necessary precautions did not affect the program, since the number of volunteers was sufficient to achieve the intended aims: the study and research objectives as well as the aims leading to an improvement of the road network by incident detection, areas likely to be particularly dangerous or new road developments.[18]

3.2 *Measures to Hinder the Re-Identification of Drivers*

The identification or re-identification of a natural person is a central criterion for qualifying personal data and constitutes the entry point for the risk of infringement of the rights and freedoms of natural persons involved in data collection. To respect these rights, data must be anonymised or at least pseudonymized.

---

[17] A Joint Technical Committee includes staff representatives and management.

[18] Below is a link to the results of the experimental phase in 2015. The new phase, currently in the process of collecting consent, already shows that a sufficient number of drivers will volunteer:

http://www.lma.ifsttar.fr/linstitut/ts2/laboratoires/lma-ifsttar/evenements-scientifiques/seminaire-s-vrai/.

3.2.1 Irreversible Anonymisation, an Objective that Is Difficult to Reach and not Always Relevant for the Purpose Intended

Based on the directive 95/46/EC, the WP29[19] defines anonymisation as "*the result of the processing of personal data in order to prevent, irreversibly, any identification*". ISO 29100 defines it more technically as the "*process by which personally identifiable information (PII) is irreversibly altered in such a way that the subject of the PII can no longer be identified directly or indirectly, whether by the person responsible for processing the information. IPI alone or in collaboration with any other party*".[20] With irreversible anonymisation, data lose qualification of personal data and therefore fall outside the scope of the legislation.[21]

Considering the multitude of re-identification techniques, the irreversibility of anonymisation is complex to implement. In addition, irreversible anonymisation is not always possible or even desirable if the purposes require authentication of the vehicle[22] and for the preservation of the exercise of the right of access to their personal data by the data subjects.[23] Moreover, road risk prevention requires collection of geolocation (accurate knowledge of crash sites), and in some cases of instantaneous speed (studies in accidentology, possible proposals for appropriate speed limitation measures,…).

Pseudonymisation is then the most appropriate method.

3.2.2 Pseudonymization, An Appropriate Method

Pseudonymisation consists of "*replacing directly identifying personal data by a non-signifying pseudonym*".[24] As these pseudonymised data are indirectly identifiable, they remain subject to the protection rules and all the collected data associated with them. Pseudonymisation is not a method of anonymisation but a security measure that reduces "*the correlation of a set of data with the original identity of a person concerned*".[25] It can thus "*reduce the risks to the data subjects concerned and help controllers and processors to meet their data protection obligations*".[26]

---

[19]The WP29 is the former grouping of European regulatory authories that has been replaced by the EDPS with the implementation of the GDPR.

[20]ISO standard 29100:2011, taken over by the opinion 05/2014 of the WP29, p. 6.

[21]Op. cit.: GDPR, recital 26.

[22]. A vehicle authentication may be required to provide a service or to provide evidence in the event of an accident.

[23]ECJ, Judgment C-553/07, 7 May 2009 College van burgemeester en wethouders van Rotterdam/M.E.E. Rijkeboer.

[24]Op. cit.: Compliance Package, p. 7.

[25]WP29, 2014.

[26]Op. cit.: GDPR, recital 28.

In view of the aims of the S_VRAI and DYMOA projects and the need for some data that may be identifiable, technical and organizational methods to pseudonymize data collected have been developed. For example:

- Identifying data needed to install the boxes to perform their maintenance and to allow users to exercise their access rights, have been placed in a correspondence table associating them with an encrypted identifier. This correspondence table, hosted in a separate medium from the operating database, is accessible only to a limited number of people who are personally liable for unauthorized disclosure. It is erased at the end of the collection period and the scientific consolidation of the data.
- Data recorded in the boxes are encrypted before transmission in the IFSTTAR server. They are then compressed before being transmitted and cut into several "packets". The reconstitutions are carried out in the server by a member of the IFSTTAR team with rights in this respect and who has signed a confidentiality agreement. In addition, asymmetric encryption keys are used, which excludes the possibility of decoding a "data packet" in case of hacking during the transfer.
- Results sent to road managers are aggregated. Besides, data on a site concerning only one driver are eliminated from the maps presented to them because managers concerned are also hierarchical managers of the drivers and could identify their agents.

3.3 Geolocation and the Principle of Minimizing the Data Collected

Since the introduction of personal data protection legislation in Europe[27] it is imperative to limit their collection to only the data necessary and proportionate to achieve the purposes is required.[28] The GDPR continues to impose this rule: "*Personal data shall be: (…) adequate, relevant and limited to what is necessary for which they are processed*".[29]

This principle is fully implemented in the S_VRAI and DYMOA projects. Not only are the data necessary to achieve each purpose collected, but file partitioning

[27]Directive 95/46, article 6-c.

[28]This even though these would not be the initial purposes for which the data were collected, but subsequent purposes authorized by law, such as scientific or historical research or research for statistical purposes. The GDPR, which also admits the subsequent use of data for purposes that are "*not incompatible*" with the original purposes [Art. 5 (1) (b)], recalls that "*appropriate guarantees for the rights and freedoms of the data subject*" must nevertheless be put in place (recital 156). Data controllers must therefore develop "*technical and organisationnel measures*" to comply with the principles imposed by the Regulation and ensure respect for the rights and freedoms enshrined in the Regulation.

[29]GDPR article 5-1-c.

measures have been defined as well as access authorization management measures by members of the research teams.[30]

3.3.1 The Incident and Geolocation

The study of the accumulation of incidents at the same place, is the main object of both projects. The location of the vehicle and its instantaneous speed are therefore essential, which may raise the question of the risk of revealing an infringement of the speed regulation (see Part I for the regulations in this respect). However, on the one hand, the teams with access to the data are members of public research and study organisations and, on the other hand, the objective does not require collection throughout a trip. This is why the data collected is only stored between 30 s before and 15 s after the case is triggered. This triggering occurs during an incident. It may also intervene during the passage through an area considered to be a risk or during the crossing of a development to be studied. This duration is considered by accidentology researchers to be sufficient to understand the course of events. A nearly similar duration was recommended by the CNIL in the Compliance Package.[31] An integrated camera films the scene for the same length of time, which helps to contextualize the reasons for the incident (e.g. vegetation hiding visibility on a traffic sign before an intersection and after a curve).[32]

This collection also provides the necessary elements for the road manager concerning areas to be taken into account to improve the safety of his network.

3.3.2 GPS Tracks and the Calculation of the Stake

In order to assess the issue represented by the incidents collected on a specific network, it is necessary to know the volume of the equipped vehicles circulating on this network, thus to note all their passages on said network. To avoid individualizing the route of each vehicle, and thus avoid tracing their drivers, the information was collected in a separate file, without reference to a particular vehicle. The necessary information here is as follows: during the duration of the experiment 'N' equipped vehicles circulated on this network and produced, together, 'N' incidents in position 'X' or position 'Y'. In addition, the data is recorded at a lower frequency than for incidents.

---

[30]IP address IP of authorised remote computers, login et password for authorised people.

[31]Op. cit.: Compiance Package p. 26.

[32]The rights of third parties are preserved since the license plates and faces of pedestrians who may be in the field of the camera are blurred.

3.3.3 The Development of a Speed Observatory

To develop an observatory of speeds, here again only the necessary data are recorded: GPS position of vehicles, instantaneous speed expressed in km/h. The combination of these data could set to fears of identification and again to reveal a violation of speed limit regulations. As only the two data cited are essential for developing a speed observatory, the pseudonymous identifiers of the boxes are automatically removed from the file when it arrives in the IFSTTAR server. The file sent to the processing team, hosted in a separate server, no longer contains identifying data.

The work carried out in these two projects is currently being transferred to a "start-up" which will develop the program more broadly with road managers by integrating the protection principles implemented. This respect is one of the fundamental requirements of the transfer, included in the public outsourcing contract, and the managers of the beneficiary company have immediately considered that this constitutes for them, in the long term, a real competitive advantage.

4 Conclusion

ICT has been developing at a high speed in recent years in vehicles, but technical and organisational protection measures are still difficult to implement. However, this protection is all the more essential since, in addition to protecting drivers' rights, certain measures will also prevent cybersecurity risks. Increase in cyber-attacks in the connected world in general, development of highly automated driving vehicle with potential risks for road user in the event of malicious intrusion into the systems, should encourage private and public sector stakeholders to collaborate to ensure the highest possible level of safety for road traffic system users. Only regulations relating to protection of personal data have been mentioned here. But a legal corpus on cybersecurity is being developed. International bodies for automotive technical regulations are also discussing these topics, which are increasingly being tackled in line with data protection rules.

Thus, personal data protection and cybersecurity converge to provide drivers with the best possible security and offer designers and producers of innovations a competitive advantage: ensuring data security will generate user confidence.

Appendix 1. S_VRAI

S_VRAI (Sauver des Vies par le Retour d'Analyse sur Incidents) This project, funded by the DSCR in a first phase (Road safety and traffic delegation) and authorised by the French data protection authority (Decision 2012-049, 16th Feb. 2012), aims at developing a system for collecting data on road incidents to provide a better understanding of risk. Data loggers were installed in 50 administrative service vehicles. The second phase, is currently being launched, is being carried out directly in partnership with local councils. About twenty service vehicles for each fleet should be equipped.

http://www.lma.ifsttar.fr/linstitut/ts2/laboratoires/lma-ifsttar/evenements-scientifiques/seminaire-s-vrai/

Appendix 2. DYMOA

DYMOA (Diagnostic d'infrastructure et dynamique du Véhicule pour les motos et les autos) Subsidized by the Road Safety Foundation and authorised by the French data protection authority (Decision n°2016-034, 11 Feb. 2016), this project aims at developing new methods for the diagnosis of road infrastructure safety and their uses, especially for motorized two-wheelers, by using data collected from IFSTTAR's recorded devices (EMMAPhones i.e. specially equipped and secured dedicated smartphones). The project aims at providing answers on behaviour of motorcyclists drivers by analyzing:

- dynamic behaviour of vehicles in real use,
- use of road infrastructures,

40 private vehicles have been equipped (30 motorcycles and 10 vehicles).

http://www.ifsttar.fr/linstitut/ts2/laboratoires/lma-ifsttar/contrats-et-projets-de-recherche-en-cours/dymoa-diagnostic-dinfrastructures-et-dynamique-du-vehicule-pour-les-motos-et-les-autos/

References

CNIL (2017). Connected vehicles and personal data

Guilbot M, Serre T, Ledoux V (2016a) Quelle protection pour les données personnelles des conducteurs? Revue TEC n°231, novembre 2016 pp 44–45

Guilbot M, Serre T, Naude C, Ledoux V (2016b) Legal conditions for implementing EDRs in public fleets of vehicles. In: 11th ITS European Congress. Glasgow, Scotland, June 6–9, 2016, 10p

Guilbot M, Serre T, Ledoux V (2017) Concilier la collecte des données de conduite individuelles et les droits fondamentaux des conducteurs. Chap. 3 dans Evaluation des politiques de sécurité routière. Nouvelles technologies, enjeux économiques et communication. Coord. par Blanchard G. et Carnis L. Ed. Le Harmattan, déc. 2017 pp 57–74

Guilbot M, Vaslin V, Arrègle E (2018) Véhicule connecté, communicant et protection des données à caractère personnel des usagers. Les Rencontres de la mobilité intelligente, ATEC, 24 janvier 2018, 13p

Real Time Driving Risk Assessment for Onboard Accident Prevention: Application to Vocal Driving Risk Assistant, ADAS, and Autonomous Driving

Johann Brunet, Pierre DA Silva Dias and Gérard Yahiaoui

Abstract Accident risk assessment is a research field that has been started in the 60's, in particular on the basis of «risk triangle» theory proposed by Frank E. Bird in 1969. This theory of risk uses the notion of «near misses» observation that leads to «safety rules» in a large number of application fields such as manufacturing, firemen, etc. Note that this approach is the opposite of what some automotive teams are currently starting (observing «everything» recorded on a big database): Most of the time nothing happens (no incident, no surprise). Risk experts do not do this. They are focused on «near misses» observation only, instead of observing the global process where at least 99.99% of data are not relevant for risk assessment. And observation of accidents is not relevant neither: accident is a rare combination of «near miss» situation and bad luck (stochastic process). Risk experts work on safety rules made to keep the situation as far as possible from near miss. In the domain of road safety, this approach has been developed mainly by road infrastructure experts in order to understand deep and deterministic reasons of near misses and propose ways of shaping roads and road signs in order to minimize it. This paper deals with presentation of results of a 15 years long collaborative research that extracted knowledge from road infrastructure experts and researchers, and put it into an on-board knowledge-based artificial intelligence in order to score driving risk dynamically and in real time. Indeed, there is no inherently bad driving behaviour and there is no inherently dangerous infrastructures, contrary to what many people think (black spots, harsh/brutal driving behaviour, etc.). It is when driving behaviour is inappropriate to driving context that driving risk appears. Driving context is described by infrastructure characteristics (geometry and functionality), by location of other users of the infrastructure (cars, trucks, bicycles, pedestrians, …), by weather conditions, etc. In this list, infrastructure takes 75% of the global weight, and then the artificial intelligence uses the digital map, and does pattern recognition on this digital map, in order to describe infrastructure context. ADAS sensors give additional inputs such as inter-distance, size of free space, time to collision, visibility measurement … Then driving risk depends dynamically on the context (infrastructure, traffic, visibility, …)

J. Brunet · P. DA Silva Dias · G. Yahiaoui (✉)
NEXYAD, 95 rue Pereire, 78100 Saint Germain en Laye, France
e-mail: gyahiaoui@nexyad.net

J. Langheim (ed.), *Electronics Components and Systems for Automotive Applications*, Lecture Notes in Mobility,
https://doi.org/10.1007/978-3-030-14156-1_23

and on the driving behaviour, through a sensor plus map fusion, made by the artificial intelligence. This artificial intelligence has been integrated inside a software API called SafetyNex that is already under deployment. This tool can alert human driver before dangerous situation, letting time to smoothly slow down and avoid potential bad surprises. Of course it can also alert the AD (Autonomous driving system) in order to automate a cautious driving behaviour (in urban areas for instance). The Artificial Intelligence estimates driving risk 20 times per second and this paper shows on real uses cases how this risk assessment anticipates potential danger. In addition, risk profiles are recorded on the cloud. We think that this risk profile recording (both for human driver and autonomous car) can be a start of comparison in terms of road safety, and a good communication vector from car manufacturers to car insurance that will need «proofs» if they ever propose a special cheap pricing. It is important as car insurers are expected to pay a part of the value brought by ADAS and autonomous vehicle because they act on risk. Of course, car insurer will accept if and only if ADAS and AD reduce accident rate and cost loss. So we conclude on the economic value of artificial intelligence in connected and intelligent cars using the cloud to exchange risk data.

Keywords Risk · Driving risk · ADAS · Autonomous driving · Scenario-based · Complexity · Validation · Adaptive · Adaptive AD · Real time driving risk computing · Car insurance · Fleet management

1 Introduction

Automotive electronics has recently experienced significant growth under the pressure of ADAS and improvement of comfort in the coskpit. The skills of the engineers who are needed to develop this type of system are both diverse and relatively new for the automotive world: electronics and computing capabilities, sensors (ultrasound, radar, lidar, camera, …), sensor fusion, signal and image processing, computer vision, artificial intelligence, …

It is therefore natural that the teams that started to develop this type of skills found themselves in charge of working on the autonomous vehicle.

However, the autonomous vehicle and the ADAS systems are two very distinct entities that require to have a quite different state of mind. Indeed, in the case of ADAS, there is a human driver inside the car: that drives, that anticipates, that does one's best to never be surprised by an emergency situation.

An autonomous driving system (AD) must support this task of anticipation and cautious driving.

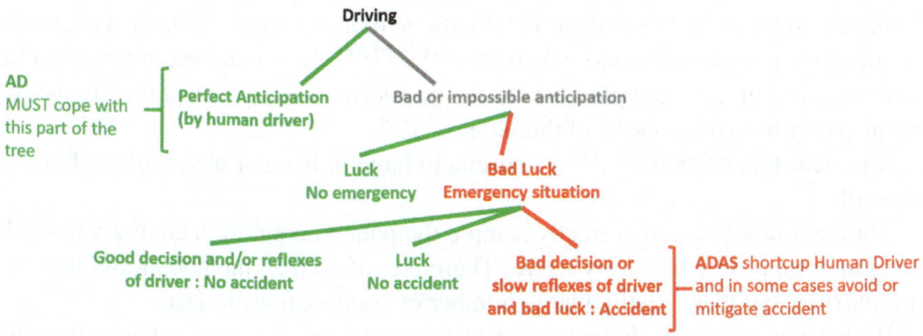

In the following chapters, we explain that the engineering approaches implemented in the context of ADAS are, in our opinion, largely insufficient in the case of autonomous vehicles, and we show how to improve these approaches.

2 The Scenarios Method: Is There an Alternative?

Most automotive engineers are familiar with the so-called "scenario" approach.[1] It is a question of listing a large number of situations of life in driving, and to ask each time what is the appropriate response of the autonomous driving system. The stakes are to build the chain of perception and understanding of the situation of life, then to decide what is the best possible response of the vehicle in every situation.

When we say "the best possible response", it obviously integrates the fact that the vehicle is put if possible not in danger. In other words, the concept of driving risk that the vehicle takes at every moment is present in this approach, but implicitly: we never need to measure this risk since by definition it is the lowest possible in each situation.

Criticism of such an approach called "feed forward" can be presented along two axes:

- the completeness of the cases: obviously, there is no evidence that between two scenarios recorded during the design of the system there is not at least one third with an intrinsic danger that would not be an interpolation of the two known cases. This difficulty of designing the feed forward chain is then found again in the validation task: how many scenarios and which ones should we validate in order to be able to affirm that the system is "working"?
- the cost of the technical means necessary to carry out the scenarios: they must be recorded in large databases of actual or simulated driving, and labeled, etc.

To try to understand what is "a lot" in the field of autonomous vehicles, just consider a commonly accepted sensor: the camera. A standard 8-bit color HD camera has 3 8-bit color planes, then a total of 24 bits per pixel. The number of pixels is about

[1]Toward a discipline of scenario-based architectural engineering, Rick KazmanS. Jeromy CarrièreSteven G. Woods, annals of software engineering, May 2000, Volume 9, Issue 1–4, pp. 5–33.

2 million. Since 2^{24} is more than 65,000, the number of images that such a camera can display on its sensor is more than $65{,}000^{2{,}000{,}000}$. To get an idea of this gigantic number, one can, for example, refer to the physicists' assessment of the number of atoms on Earth that would be of the order of 10^{50}.

It is clear that $65{,}000^{2{,}000{,}000}$ is extremely large in front of the number of atoms on Earth.

Thinking that you can correctly sample the phenomenon by realizing a few billion km seems to be then very naïve. There are, of course, methods to reduce this complexity,[2] but there is still a large number of combinations to test.

We have proposed an alternative, or even complementary, approach which would be to explicitly measure at each moment the risk of driving that the driver (human or not) is taking.[3]

And for that, we have a lot of information. Indeed, when a human driver must pass his driver's license, in order not to be dangerous for himself and for others, this driver must learn a number of rules of cautiousness.

These precautionary rules deal mainly with the following topics:

- the shape, and the functionality of the road infrastructures
- the nature, position, and attitude of other road users
- weather conditions.

It is on this basis that we worked for fifteen years in order to compile a collection of this (worldwide) expertise. Participation in numerous collaborative research programs, including in France PREDIT[4] programs on road safety, allowed us to understand the definition and estimation of driving risk related to these rules. In particular, we have found that unsuitable driving behavior in the infrastructure carries 75% of the risk of accidents, which is not well known in the automotive sector traditionally interested in … mobile (and not in road infrastructures).

3 A Real-Time Driving Risk Calculation System

We have been able to develop and validate an effective real-time driving risk assessment system. This system is currently being deployed by automobile manufacturers to alert the human driver in advance, when the calculated risk is high, allowing time to slow down, thus avoiding being in an emergency situation.

---

[2]Building a relevant validation database for camera-based adas, Gérard Yahiaoui & Pierre Da Silva Dias, 1st of june 2016, Conference Autonomous Vehicle Test and Development 2016, Stuttgart.

[3]Functional Safety: On-Board computing of Risk of Accident, Grégoire Julien, Pierre Da Silva Dias, Gérard Yahiaoui, 2 sept. 2016, Advanced Microsystems for Automotive Applications 2016 Part of the series «Lecture Notes in Mobility» pp. 175–180, Springer.

[4]Evaluation du risque de sortie de route pour l'aide à la conduite ou le diagnostic d'infrastructure, Johann Brunet, Pierre Da Silva Dias, Gérard Yahiaoui, 3 mai 2010, Conférence PRAC 2010 (Prévention des Risques et Aide à la Conduite)—Session caractérisation du risque routier.

The studies we conducted on this subject show that it is possible to avoid 20% of road accidents, which is not negligible.[5]

In the case of the autonomous vehicle, it can be said that such a calculation allows the Artificial Intelligence of autonomous driving to be "aware of the risk it takes". If the risk is high, even if we do not know a priori why, we can then very well decide to apply heuristics of cautiousness such as for example: "if risk too high, then slow down".

This makes it possible to envisage implementing an adaptive schema of the following type:

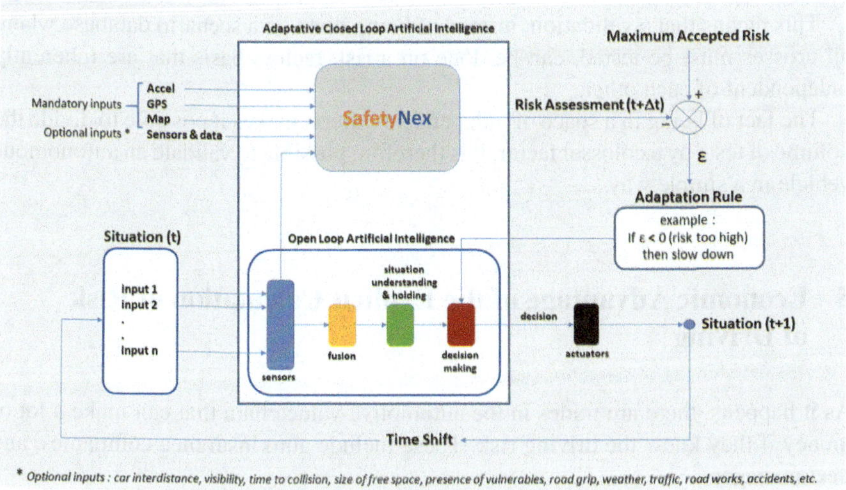

Notes:

By nature, an adaptive system is more robust than a feed forward system. It is possible and even desirable to combine these two approaches to adapt the current chain when we come across a case that was not planned by the engineers.

One could even imagine doing AD by learning: applying reinforcement learning, reinforcing decisions that lead to low risk, and penalizing decisions that lead to high risk.

[5]Aide embarquée à la conduite sûre et à la prévention du risque de conduite, Adil el Aboubi, Johann Brunet, Pierre Da Silva Dias, Gérard Yahiaoui, 25–26 Juin 2018, Congrès SIE «Regards Croisés sur le véhicule autonome», Paris 25 juin 2018, SIA.

4 Advantage of the Explicit Calculation of the Risk in Terms of Validation

Validation is the fact that the risk remains permanently below its maximum accepted value.

It's a simple way to validate.

Moreover, the risk, in its structure of calculation, is not natively a process of compensation: when one makes a big error of driving relating to 1 risk factor, one can not be in safety on the pretext that one is perfect with respect to all other risk factors.

This means that a validation, instead of being done on a scenario database where all crosses must be tested, can be done on a risk factors basis that are inherently independent of each other.

The fact of being in a space of independent criteria makes it possible to divide the volume of tests by a colossal factor. It is therefore possible to validate an autonomous vehicle in a simple way.

5 Economic Advantage of the Explicit Calculation of Risk of Driving

As it happens, there are trades in the automotive value chain that can make a lot of money if they know the driving risk. These include auto insurance companies, and fleet managers.

If the autonomous vehicle, or even partial systems of driving delegation, can provide these interlocutors risk values that they are likely to use, we open the way to an opportunity to share the "road safety" value in form of revenue sharing.

Improving ITS-G5 Cybersecurity Features Starting from Hacking IEEE 802.11p V2X Communications Through Low-Cost SDR Devices

Vincenzo Di Massa and Samuele Foni

Abstract This paper analyzes V2X and IoT autonomous vehicles cybersecurity risks. Security assessments are a demanding activity because the auditor must gain domain specific knowledge during the assessment itself. One important aspect of investigation is the potential for the system users to inadvertently compromise security measures or privacy restrictions on data access. The autonomous driving (AD) domain is one of the first examples of application in which safety critical decisions are taken collaboratively by system users and Artificial Intelligence algorithms. Thus while the AD system must help the user driving and the driver/passenger to be able to control the autonomous car overall behavior, the on board security system must avoid rogue users from compromising the travel safety and availability first while also guaranteeing the user privacy. In this context security and safety can mandate seemingly contradicting requirements and also usability can be impaired by security measures. AD projects are trying to evolve risk assessment methodologies in order to deal with such scenarios. Moreover the introduction of IoT technologies into the AD context (e.g. the Autopilot project) is an important vulnerability that has to be taken into account. This paper describes an example in which a standard framework, designed for Industrial Automation Control Systems to V2X and IoT autonomous driving, is employed while keeping into account the AD and IoT requirements.

Keywords V2V communication · Connected car · Cybersecurity · 802.11p · Artificial Intelligence · Autopilot · IoT · IEC 62443

V. Di Massa (✉)
Cybersecurity Expert and Cybersecurity Authority ICS, Thales, Italy
e-mail: vincenzo.dimassa@thalesgroup.com

S. Foni
Cybersecurity Expert, Thales, Italy
e-mail: samuele.foni@thalesgroup.com

J. Langheim (ed.), *Electronics Components and Systems for Automotive Applications*, Lecture Notes in Mobility,
https://doi.org/10.1007/978-3-030-14156-1_24

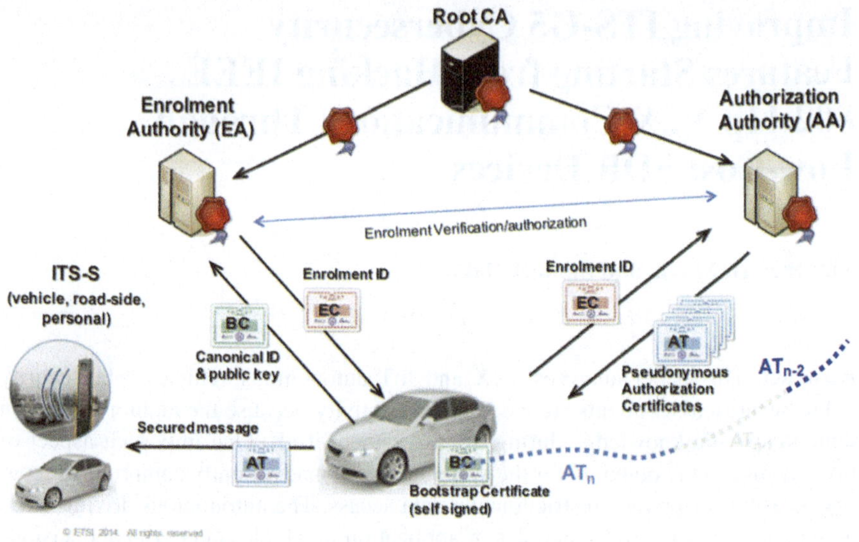

Fig. 1 ETSI security reference model

1 Introduction

AUTOPILOT is an Horizon 2020 project that intends to develop Autonomous Driving (AD) use cases by augmenting AD with Internet of Things (IoT) technologies and infrastructure.

With the increasing adoption of IoT, new security challenges need to be addressed as the threat of attacks is moving from the digital to the physical world, leading to even more severe safety implications.

Security is critical to IoT's adoption, especially in AUTOPILOT; because we want to make sure we can "trust" data flowing between sensors, actuators, rules engines and other connected components of our architecture. Furthermore, when IoT's devices are used for AD (Autonomous Driving) functionalities, as addressed by the AUTOPILOT project, security aspects must be stressed because matters of safety and national security may be at stake. Autonomous vehicles, if used as a weapon, would cause substantial harm to people and societies (Fig. 1).

Stakeholders can address these IoT security challenges around the following principles from mainstream cybersecurity standards, but not all of them are equally useful in this context. In AUTOPILOT, after a detailed analysis, the ISA IEC 62334 series of standards has been selected to guide cybersecurity processes.

The existing security technologies and methodologies need to evolve from their current status to address all the new IoT and AD security issues. In AUTOPILOT no new security components has been designed, but effort has been put into understanding how to address the IoT enhanced AD with state of the art technology and processes.

From a technological point of view Autopilot is an ETSI Intelligent Transport System (ITS) augmented by IoT.

The Security, Privacy and Data Models for this project can be seen as an evolution of ITS methodologies that integrate IoT measures, thus the AUTOPILOT risks, threats, assets and stakeholders are a superset of those of ITS and IoT.

In the AUTOPILOT context, the traditional Confidentiality, Integrity, Non-repudiation, Availability, Authenticity and Accountability security objectives must all come after safety and help to ensure safety.

Safety, even if not strictly in the scope of this analysis, has major weight in deciding the risk ranking, the mitigations and the requirements for AUTOPILOT.

As perfect security does not exist, the design of security features in a safety critical environment always puts safety as the top priority. ETSI, as in the ETSI TR 102 893, addresses the core threats, risks and vulnerabilities for ITS-G5.

2 Red Team Testing

Our research goal within the first validation phase of the AUTOPILOT project has been focused mainly on the identification of the cyber risks that the ITS-G5 infrastructure could be exposed to. In order to thoroughly investigate this aspect we have chosen to analyze the ETSI G5 standards from the point of view of a potential attacker, using tools easily available on the market, exploiting a laboratory environment offered by the project in which all the recommended cybersecurity policies had not yet been applied. In this way we have been able to demonstrate the maximum capabilities of the cyber-attacks that could be made against an ITS-G5 infrastructure not implementing all the recommended security features such as cryptography. Once a good level of awareness in this topic has been obtained, the next phase of the project will have the objective of making use of all the technical details collected to achieve the effective securing of the ITS-G5 testing environment deployed within AUTOPILOT. The attacks described in Chaps. 2 and 3 demonstrate the feasibility of such an attack with just limited time and resources. Chap. 4 describes how the use of a standard risk analysis process predicted the consequences of the simulated attacks and also provided mitigations.

In order to carry out the survey from the point of view of an effective attacker, we have chosen to resort to the use of a red team-oriented analysis approach. Namely, a red team is an independent group of cybersecurity practitioners that performs cyber-attacks on an infrastructure with the aim to study its weaknesses and to improve its effectiveness, detecting and evaluating threats and vulnerabilities form an attacker-like perspective. What distinguishes a red team from a group of attackers are: customer consent and ethics. The advantage of using such strategy of test is to provide a more realistic picture of the security readiness than exercises, role playing, or standard assessments in general.

In this context, this line of attack has been used for technical research purposes. In red teaming approach, ethical hackers or white hats are responsible for system evaluation and penetration with limited (or without) any granted access to internal resources of an information security system or network.

2.1 The SDR: The Tools of the Trade

The ETSI ITS-G5 infrastructure is mainly based on the use of the IEEE 802.11p V2X communication standard, otherwise known as WAVE (Wireless Access Vehicle Environment), in the licensed band of 5.9 GHz (5.85–5.925 GHz). In addition, others two support interfaces are involved within the AUTOPILOT testbed deployed, such as: LTE, for communications with the command and control center, and the IEEE 802.15.4 standard, for the gathering of information from the scattered sensors on the roadside. In order to control all these interfaces, taking advantage of the maximum flexibility offered by a single hardware component, we decided to make use of Software-Defined Radios (SDR) which are easily available on the market.

The advent of inexpensive SDRs has redefined the wireless hacking landscape. SDRs are systems where components that have been traditionally implemented in hardware (e.g. mixers, filters, amplifiers, modulators/demodulators, detectors, etc.) are instead implemented by means of software elements. At the moment there are several types of SDR devices to which we can refer to; in our case, we based our choice on the analysis of the following five features, finding the one that best suited into the AUTOPILOT operating context:

- *Sample Rate/Bandwidth*. The sample rate defines the maximum bandwidth that we are able to view simultaneously.
- *Dynamic Range/ADC Resolution*. Higher ADC resolutions let us view high and low energy signals together and observe smaller differences in the signal. A typical resolution for most signals is done with the aid of 8 bits, but SDRs with 16 bits of resolution are still available on the market.
- *Transmit Capability*. Some SDRs allow us to transmit and receive simultaneously (full duplex), but others only allow half duplex transmit capabilities.
- *Tuner Range*. The tuner range determines what frequencies we are able to receive.
- *Price*. A too expensive SDR may not adequately represent the technology used by an effective attacker.

After these considerations, our choice is the HackRF One SDR, with the aid of a logarithmic antenna able to operate up to 6 GHz, in order to adequately overspread the operating frequencies of the IEEE 802.11p standard. To properly manage the software side we employed GNURadio, which is an open source development toolkit for programming SDRs. GNURadio Companion makes use of a graphical interface to develop the wanted behavior, generating a series of logic blocks connected to each other, called flowgraph. At the compilation time, these blocks are automatically converted into Python code and used to control the HackRF One. Finally, the self-generated code was subsequently modified by hand in order to better fill our needs (Fig. 2).

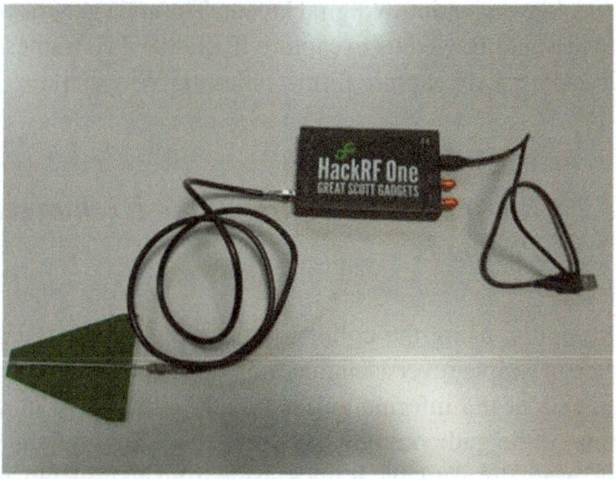

Fig. 2 The HackRF One SDR with the logarithmic antenna used in this context

3 Attack Perspective and Work Findings

In order to follow a red team approach on the prototypal environment of AUTOPILOT, not yet implementing all the appropriate cybersecurity ETSI G5 features, we performed the test activities without full awareness of the actual development status of the G5 infrastructure. The research work of the attack vectors on architecture has been structured into three main phases:

- Information Gathering.
- Reverse Engineering of the exchanged messages payloads.
- Exploiting ITS-G5 V2X communications.

3.1 *Information Gathering*

First of all, an inspection was carried out on the AUTOPILOT site, with the aim to attempt to understand what kind of technologies were actually in use, to locate as points of interest the active Road Side Units (RSU) on the route, at the pedestrian crossings and at traffic light stations. Subsequently, several sniffing sessions were carried out, taking care to keep track of all the visual variations verifiable on the environment, such as: AD car passing, mapping of the active lights on the vehicles, recording of a braking event, proximity to a green/yellow or red traffic light, etc.

The scanning for the IEEE 802.11p signals to be captured during the sniffing phase was made through the use of the open source tool Gqrx (GNU Radio and Qt graphical toolkit), which is an excellent framework for searching the analog signals within the frequency range covered by the SDRs, through a fairly intuitive graphical interface. Once the analog signals related to the exchange of IEEE 802.11p messages were identified, a module was developed by means of GNURadio for their reception and conversion into a digital layout. The code generated and executed on the HackRF One SDR allowed us to capture IEEE 802.11p frames and to collect them in various PCAP (packet capture format) files via Wireshark.

3.2 *Reverse Engineering of the Exchanged Messages Payloads*

Secondly, using the collection of packet captures from the information gathering step, a simple reverse engineering on the protocol was carried out. In order to be able to extract the information that could allow us to understand the meaning of some bits of the gathered payloads, we used temporal and spatial correlation techniques between the available frames, aside from an in-depth analysis of the ASN.1 (Abstract Syntax Notation One) code of the ETSI ITS-G5 standard documentation, and above the receiving of some indirect advices through the use of simple social engineering operations.

At the end of the investigation, a Python script was created for the extraction and interpretation of the fields of interest from an IEEE 802.11p frame. At this stage we were able to correctly derive the message structure provided by the ITS-G5 standard: Simple GeoNetworking, CAM, DENM, SPATEM and MAPEM packets.

3.3 *Exploiting ITS-G5 V2X Communications*

Finally, we moved on to the active phase, writing a new Python script using the GNURadio libraries to set up our HackRF One to transmit messages on the IEEE 802.11p channel. By simply changing the packet timestamps, the sender/receiver addresses and few other bits, we were able to reply and tamper any type of message on the channel (e.g. Simple GeoNetworking, CAM, DENM, SPATEM and MAPEM) and to have them correctly interpreted by the RSU and the car On Board Units (OBU) devices.

Potentially, at this point, an attacker would be able to carry out AD cars flooding attacks, generate fictitious traffic light signals, report a false presence of pedestrians within the road, set unfit speed limits, realize OBU and/or RSU attacks spoofing, and so on.

Becoming aware of this result, we can move on to the introduction of the cyber-security remediation to be applied to the ITS-G5 infrastructure, ensuring to make attacks of this type unfeasible in the next future.

4 Results

Industrial Automation Control Systems (IACS) standards are already employed in the field as guidance for the development of transportation systems like railways and tramways. An IACS is very different from an IT infrastructure for many reasons including:

Category	Information technology system	Industrial control system
Performance Requirements	Non-real-time Response must be consistent High throughput is demanded High delay and jitter may be acceptable Less critical emergency interaction Tightly restricted access control can be implemented to the degree necessary for security	Real-time Response is time-critical Modest throughput is acceptable High delay and/or jitter is not acceptable Response to human and other emergency interaction is critical Access to ICS should be strictly controlled, but should not hamper or interfere with human-machine interaction
Availability (Reliability) Requirements	Responses such as rebooting are acceptable Availability deficiencies can often be tolerated, depending on the system's operational requirements	Responses such as rebooting may not be acceptable because of process availability requirements Availability requirements may necessitate redundant systems Outages must be planned and scheduled days/weeks in advance High availability requires exhaustive pre-deployment testing
Risk Management Requirements	Manage data Data confidentiality and integrity is paramount Fault tolerance is less important - momentary downtime is not a major risk Major risk impact is delay of business operations	Control physical world Human safety is paramount, followed by protection of the process Fault tolerance is essential, even momentary downtime may not be acceptable Major risk impacts are regulatory non-compliance, environmental impacts, loss of life, equipment, or production
System Operation	Systems are designed for use with typical operating systems Upgrades are straightforward with the availability of automated deployment tools	Differing and possibly proprietary operating systems, often without security capabilities built in Software changes must be carefully made, usually by software vendors, because of the specialized control algorithms and perhaps modified hardware and software involved

(continued)

(continued)

Category	Information technology system	Industrial control system
Resource Constraints	Systems are specified with enough resources to support the addition of third-party applications such as security solutions	Systems are designed to support the intended industrial process and may not have enough memory and computing resources to support the addition of security capabilities

A transportation system has many points in common with an IACS, but also some differences.

Similarities include the required level of security and availability and the impact of the system on the local economy. Both transportation systems and IACS can be seen as critical infrastructures, most of the time being operated by a single company.

What makes them different is mainly the distributed and public nature of transportation systems.

This difference has deep implications on security and privacy requirements: while it is often possible to physically protect an IACS from attackers, by means of physical access control and network segregation, it is probably impossible to provide the same level of physical protection to a transport system.

Following this consideration, the IACS security standards, such as ISA/IES 62443, are useful to provide guidance for some parts of the systems, especially for Control Centres. Moreover IACS offers a good approach in which security is analysed in a context where a failure can have very high costs both in terms of human lives and money.

Given that ISA/IEC 62443 is technology-independent, Autopilot was able to use it when analysing requirements, even if Autopilot is not designing an IACS.

4.1 The "Architecture" of the ISA IEC 62443 Standard

The ISA IEC 62443 is a multipart standard whose parts can be grouped in 4 layers. From the bottom to the top (Fig. 3):

1. *Component level*
   Development Lifecycle and the single component requirements are described in this layer. Component suppliers use this part of the standard to provide the appropriate security quality to their products.
2. *System level*
   The risk analysis process and the system level requirements are described in this layer. System integrators use this level to design the appropriate mitigations and the overall system architecture that allow a desired security level as identified during the risk analysis process.
3. *Policies and Procedures Level*
   The policies and procedures for asset owners and operators are described at this layer.

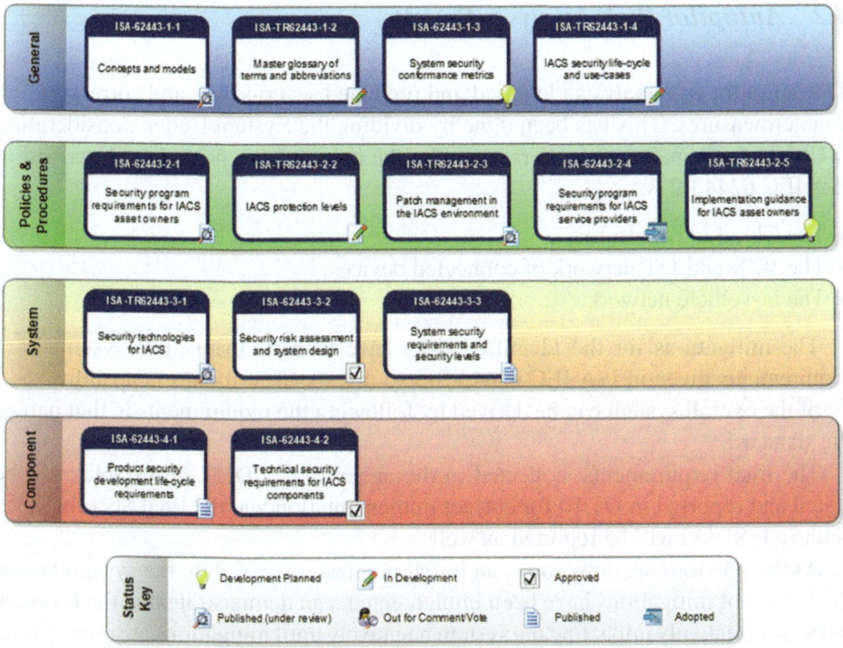

Fig. 3 The “architecture” of the ISA IEC 62443 standard

Asset owners and operators use this part to define their own processes similarly to what the ISO 2700x standards require.

4. *General level*
   The general definitions and concepts are described in this layer.
   Other parts of the standard reference this part.

ISA IEC 62443 defines security levels as a tool to describe the system’s resistance level against different attackers ranging from unexperienced “script-kids” to government funded spy agencies.

The ISA IEC 62443 also defines 7 Foundational Requirements (FR) that will be used in Sect. 4:

- FR 1—Identification & authentication control;
- FR 2—Use control;
- FR 3—System integrity;
- FR 4—Data confidentiality;
- FR 5—Restricted data flow;
- FR 6—Timely response to events;
- FR 7—Resource availability.

4.2 Autopilot Risk Analysis Results

The autopilot risk analysis identified and predicted several risks and corresponding countermeasures. This has been done by dividing the System Under Consideration (SUC) in three main zones, as required by the risk analysis procedure described in ISA IEC 62443-2-3:

- The Cloud IoT platform,
- The V2X and IoT network of connected devices,
- The in-vehicle network.

The mitigations for the identified risks have then be mapped to system level requirements given in ISA IEC 62443-3-3 so that the security level capability (SL-C) of the overall system can be derived by following the requirements in that part of the standard.

The final requirements, collected in the annex of the D1.9 deliverable, will be tested and reported in D1.10 for correct implementation and the final Security level achieved (SL-A) will be reported as well.

As the previous sections show, an initial pen-test, executed on the system before the full set of mitigations have been implemented, can demonstrate that the foreseen risks are effectively impacting the system negatively until mitigations are put in place.

The use of the ISA IEC 62443 has been beneficial in both providing grounded guidance to the risk analysis process and in deriving results and mitigations that can be easily tested, understood and compared. In fact the lowest security level (SL-1) defined by the standard would have been enough to prevent such types of attacks. Autopilot risk analysis requires a security level equal to, at minimum, SL-2.

The manufacturer's authorised representative in the EU is Springer Nature Customer Service Centre GmbH, Europaplatz 3, 69115 Heidelberg, Germany. If you have any concerns regarding our products, please contact ProductSafety@springernature.com

Printed and bound by CPI Group (UK) Ltd, Croydon, CR0 4YY

30/09/2026

02225643-0009